T0361227

ROUTLEDGE LIBRARY EDITIONS: HISTORY
AND PHILOSOPHY OF SCIENCE

SCIENCE, INDUSTRY AND SOCIETY

SCIENCE, INDUSTRY AND SOCIETY

Studies in the Sociology of Science

By

STEPHEN COTGROVE AND STEVEN BOX

Volume 7

Routledge
Taylor & Francis Group

LONDON AND NEW YORK

First published in 1970

This edition first published in 2009 by
Routledge
2 Park Square, Milton Park, Abingdon, Oxon, OX14 4RN

Simultaneously published in the USA and Canada
by Routledge
270 Madison Avenue, New York, NY 10016

*Routledge is an imprint of the Taylor & Francis Group, an informa
business*

British Library Cataloguing in Publication Data
A catalogue record for this book is available from the British Library

Library of Congress Cataloging in Publication Data
A catalog record for this book has been requested

ISBN 10: 0-415-42029-6 (Set)
ISBN 10: 0-415-47446-9 (Volume 7)

ISBN 13: 978-0-415-42029-7 (Set)
ISBN 13: 978-0-415-47446-7 (Volume 7)

Publisher's Note
The Publisher has gone to great lengths to ensure the quality of this
reprint but points out that some imperfections in the original copies
may be apparent.

Disclaimer
The Publishers have made every effort to trace copyright holders and
would welcome correspondence from those they have been unable to
contact.

Science, Industry and Society

studies in the sociology of science

by

STEPHEN COTGROVE

Professor of Sociology
Bath University of Technology

and

STEVEN BOX

Lecturer in Sociology
University of Kent at Canterbury

London
GEORGE ALLEN & UNWIN LTD
Ruskin House   Museum Street

© *George Allen & Unwin Ltd., 1970*

SBN 04 301021 0 *cased*
04 301022 9 *paper*

Printed in Great Britain at the Pitman Press Bath

Preface

Two major changes have characterised science in the twentieth century. Firstly, there has been its extremely rapid growth. Indeed most of the scientists who have ever lived are alive today. Secondly—and central to the theme of this book—science is no longer mainly an academic activity carried on in universities. Industry will soon be the largest single employer of scientists.

Such changes have generated a series of problems and given rise to much anxiety. There is growing concern lest the bureaucratisation of science shall threaten creativity. A substantial literature in America argues that scientists in industry experience particular strains and conflicts, centring round the limitations on their autonomy and restrictions on publication. And, during the period of the researches reported in this study, there has been a growing concern about the failure of industry to recruit the 'best' graduates, and the need to modify university courses to bring them closer to the needs of industry. It is such issues that this book is about.

We found less evidence of strain and conflict among industrial scientists than we had been led to expect. But this we discovered was partly explained by the kinds of scientists recruited. Indeed, a major contribution of this study is an attempt to distinguish between types of scientist; between academics on the one hand dedicated to the advancement of knowledge, and 'organisational' scientists on the other hand, who are more likely to find rewarding the application of science in the development of new products. It is the academic scientist who experiences most strains and conflicts in industry, but industry recruits relatively fewer of these, and is, in any case, more likely to employ them on basic research.

But if industry recruits relatively fewer of the academic scientists, it also recruits few of the most 'able'. However, we would put much less emphasis, than many recent reports, on the influence of the universities in producing 'academics' who, when the time comes to look for a job, find industry uncongenial. Our evidence indicates that it may be because academic research is seen to be more attractive than industry to the good honours graduate, that some undergraduates *become* academics. It is identification with a future occupation, and not the injection of academic values, which is the key. In short, just as the medical student *becomes* a doctor at medical school, so the scientist who is attracted to university research is more likely to become an academic. The crucial problem is, how can industry be made more attractive as a career for scientists? And we would argue that this will involve more than a public relations job.

If we found less evidence of strain among industrial scientists than we expected, we *did* find considerable evidence to suggest that industry is failing to use the skills and capacities of many of its scientists to the full—a major factor in the failure of industry to attract the best graduates. Moreover, we found considerable differences between laboratories. Some have succeeded in accommodating scientists in such a way that they are well satisfied. Others, partly because of the kinds of scientists they have recruited, but partly, too, because of their managerial policies, may well be failing to do their best not only for their scientists but also for themselves.

This book is written in the hope that it will make a constructive contribution to the problems of relating the two worlds of science and industry. But it is also offered as a contribution to sociology. The more technical aspects, however, have been dealt with in extended chapter-end notes and appendices. Thus, after a preliminary chapter, which places the problem in the broad context of the growth of science and its increasing application to industry and growing bureaucratisation, Chapter 2 explores more deeply the nature of science as a social system, the meaning of science to the scientist, and the different types of scientist and his relation to the values of science. Chapters 3 and 4 examine the making of a scientist and throw new light on the sociology of occupational socialisation and selection. Chapters 5, 6 and 7 explore the interactions between individuals and organisation in the industrial research setting. Chapters 5 and 6 investigate the career problems of scientists in industrial research, including problems of role strain and conflict. Chapter 7 probes the relative significance of individual and organisational variables in role performance, as measured by the productiveness of scientists.

The book is the joint effort of the authors, but Chapters 3 and 4 were contributed mainly by Steven Box, who studied the socialisation of scientists in a doctoral thesis for the University of London. We are also indebted to Julienne Ford for her contribution to Chapter 4.

We should like to thank the Social Science Research Council (then DSIR) for a grant which made possible the field research reported in this book, the Polytechnic, Regent Street, for accommodating the research unit and providing data processing facilities during this phase, and the Editors of *Sociology* and *Technology and Society* for permission to reproduce in Chapters 3 and 7 previously published material. We are particularly grateful to the many scientists working on the bench, or in the direction of research, who gave so generously of their time and received us in so friendly and helpful a fashion. Without the help of all of these, this study could not have been written. Lastly, we wish to thank Mrs Avril Fordham for the expert way in which she typed a difficult manuscript for publication.

STEPHEN COTGROVE
Bath University of Technology

STEVEN BOX
University of Kent at Canterbury

July, 1969

Contents

Chapter 1

Science and Industry

Between 80 and 90 per cent of all scientists who have ever lived are alive today. And nearly 90 per cent of the current stock of scientific knowledge has been discovered within the last fifty years. Indeed, ever since the eighteenth century, the number of scientists, of publications and of abstracting journals have been doubling every ten to fifteen years.[1]

Size alone will inevitably raise problems. And there are a number of ways in which Big Science is likely to differ from Little Science.[2] The publications explosion and the growing problem of communication is an obvious example. One man can monitor only a small fraction of the 30,000 journals. Indeed, it is estimated that the half-life of papers in physics in the USA, is now about two and a half years.

THE CHANGING ROLES OF SCIENTISTS

But it is not with the problems of size with which we are mainly concerned here. The growth of Big Science has been accompanied by changes in the sources of scientific patronage. Consequently, there have been major changes in the roles which scientists perform. Science is no longer mainly an academic activity. Particularly since the first World War, governments and industry have become the big employers. Industry will soon be the main single employer of scientists. Already in the United Kingdom 28 per cent of all scientists are employed in industry,[3] and about 70 per cent of all chemists engaged in research and development work are to be found in industrial laboratories. And it is this increasing industrial application, and the resulting growing demand for scientists to work in industry, which is generating a series of problems and difficulties. Such problems stem in part from the essential

1

differences between universities and university science, and industrial science. The objectives of universities and industry are basically different. The universities are concerned with the advancement and dissemination of knowledge. It is not surprising, therefore, that those who work in university science attach the greatest importance to free inquiry and to the disinterested pursuit of knowledge, regardless of any possible practical application. By contrast, industry is interested in scientific knowledge only in so far as it can be applied to the development of new or improved products.

CAREERS: ACADEMIC VERSUS INDUSTRIAL

It is hardly surprising then, that, when industry comes to employ science graduates fresh from university, there are complaints about their lack of awareness of industry's needs—they are not cost-conscious; are preoccupied with academic scientific solutions rather than practical ones, and lack a sense of urgency. More recently, there has been growing anxiety over the reluctance of the majority of the most highly qualified graduates to choose a career in industry. The report of the Swann Committee[4] drew attention to the fact that 79 per cent of the 'firsts' in chemistry went on to research or further academic study. For all disciplines, above average proportions of all 'firsts' and 'upper seconds' are found in universities, and a below average proportion of scientists (*not* technologists) with good honours degrees choose industry.[5] Subsequently, a high proportion (40 per cent) of those with higher degrees in chemistry gain employment in higher education and research, compared with 18 per cent in industry, although the proportions recruited to industry in other science subjects are higher.[6]

Now, one point of very considerable significance to which the Swann Report drew attention is the fact that 52 per cent of those who achieved 'firsts' or 'upper seconds' had not made up their minds about their future jobs on entering university, while 13 per cent were aiming at a university career. But before taking finals, 47 per cent had already decided on a university career, while the proportions making other choices had remained roughly constant. The inference drawn by the Swann Committee is that these findings 'confirm our views on the importance of the role of education in influencing patterns of employment'. But the problem is clearly complex. There are other factors at work besides the influence of education. Both good honours and pass degree students are exposed to the influence of education. And although the Swann Report shows marked differences in the characteristics of jobs as seen by scientists in the various sectors, this information by itself is not sufficient to explain the reasons for choice. For example, 79 per cent of scientists in universities, compared with 21 per cent in industry, saw their jobs as providing opportunities for intellectual development.[7] But this does not tell us how *important* it is to such scientists to enjoy such opportunities, and therefore how much such factors influenced their choice of job. In short, as Swann recognises, we need more information on the motivations of scientists. As the report emphasises, we know little about the complex process of occupational choice at this level, and it is to an exploration of this process that we turn in Chapters 3 and 4.

But anxieties about present trends are not confined to those who are alarmed about the reluctance of able graduates to enter industry, and the unsuitability of undergraduate courses for future industrial employment. There are others who are equally alarmed about the possible implications for science of the increasing employment of scientists in large scale organisations.

THE BUREAUCRATISATION OF SCIENCE

From Weber onwards, many leading sociologists have looked on the growth and proliferation of bureaucracies with despair and disenchantment, some writers even attributing all the 'ills' of modern civilisation to the increased size and scope of organisations. The resulting increase in the division of labour, epitomised in work on the assembly-line, results in the growing alienation of man, expressed in such terms as powerlessness, meaninglessness, isolation and self-estrangement[8]. Thus man is prevented from realising his full potentialities through self-actualisation in work.[9] Moreover, outside work, man has been increasingly separated from membership in voluntary associations and communities, and is left isolated and

3

unprotected against manipulation in the atomised world of 'mass society'.[10] Nor are these dominant themes confined to some sociological writers—they can also be readily seen as influential lines of thought in the visual arts, particularly films and theatre, and also in literature, notably the 'existentialist' novel of western Europe.

The bureaucratisation of science, it is argued, is now facing scientists with threats to their autonomy, to loss of control over the goals and methods of research, and loss of control over the products of their intellectual activity, and with a consequent loss of meaning in their daily lives. In short, those who have been among the most autonomous, whose lives have been among the most meaningful and have provided the greatest opportunities for self-actualisation, these too are threatened with alienation—with separation from control over the processes and products of their work.[11]

Much of the earlier discussion of this theme has centred around the assertion that there are potential tensions and conflicts wherever professionals are employed in bureaucratic-ally-structured organisations.[12] Any organisation is faced with the basic problem of allocating specific areas of activity to individuals and of co-ordinating the activities of large numbers in the pursuit of the goals of the organisation. This involves, above all, a chain of command, the exercise of control by some individuals over others, a hierarchy of authority. And if we forget any perjorative overtones which may be attached to the terms 'bureaucracy', this is essentially what any organisation of individuals requires—the allocation of roles and their co-ordination through a set of rules governing spheres of competence and procedures.

Now whatever ambiguities[13] there may be attached to the concept of a professional, there is general agreement that a professional has acquired knowledge and skills through a lengthy training, and in this sense is an expert in his field. It was natural, therefore, for the earlier writers on this problem to treat scientists as professionals.[14] The essential conflict between professionals and organisations centres around the distinction between professional and bureaucratic authority. The authority of the administrator flows from his position in the hierarchy. But the authority of the professional rests on his expertise. Employment in an organisation therefore, represents a potential threat to the autonomy of the professional to exercise his expertise, subject only to the judgement of his peers who are alone competent to assess his performance.

These generalisations can be illustrated from a fairly extensive literature, dating at least from Drucker's[15] observations in 1952, that industrial corporations need to make considerable changes in their traditional structures if they are to utilise the recent expansion in professionals employed in industry. More specifically, he argued that the administrative criteria for promotion could not be applied to professionals whose orientation was towards others with technical competence; that recognition by the organisation was not adequate for professionals who particularly sought extra-organisational professional recognition; that company goals and professional goals would often clash; that personnel practices were not applicable to professionals; and finally, that industry was particularly prone to under-employ professional capacities either because management had no proper grasp of scientists' potential contributions, or because management merely employed scientists for window-display purposes.[16]

Moore and Renck,[17] after examining work dissatisfactions experienced by 587 scientists and engineers, came to the conclusion that 'it can be stated *categorically* . . . that the chronic dissatisfaction of *professional* employees emerges out of a fundamental conflict which exists between the expectations and values of professional employees and the opportunities which they need to realise their ambitions in the industrial setting'. Furthermore, a 'number of complaints . . . suggest that the "chain of command" which typifies industrial organisations may be a source of considerable tension in the engineering departments and research laboratories. Any job-oriented person seeks recognition for himself . . . for the merit of his work'.

As a further example of the professional versus non-professional conflict Burns and Stalker,[18] suggested that scientists, compared with other longer established employees, claimed more recognition, higher status, more control, and more independence of the authority of management. These claims were partially responsible for the political and status struggle between departments, since they were seen by non-professionals as threats to existing social relationships and authority hierarchies. The resultant struggles prevented the electronics firms in this study from adopting an 'organic' organisational structure which the authors considered more suitable for firms needing to innovate.

Orth[19] too, when discussing the optimum climate for industrial research, reveals the same theoretical assumptions. 'Professional training in itself, whether it be in medicine, chemistry, or engineering, appears to predispose those who go through it to unhappiness or rebellion when faced with the administrative process as it exists in most organisations. Scientists and engineers *cannot* or *will* not . . . operate at the peak of their creative potential in an atmosphere that puts pressure on them to conform to organisational requirements which they do not understand or believe necessary.'

The most comprehensive synthesis of the relevant (mainly American) studies was undertaken by Kornhauser,[20] who summarised the conflicts between science, as a system of professional values, and industry, with its emphasis on economic and administrative values, into four main groups. Firstly, there are conflicts over goals. Scientists, Kornhauser argues, want to work as near to fundamental research as possible, to make a significant contribution to science, or at least work on the frontiers of it. Industrial management, on the other hand, wants scientists to concentrate only on those problems where results would be of benefit to the company. Where scientists are allowed to pursue other more fundamental research, their security is always precarious and project termination is endemic, often without adequate consultation. This leads to a second area of conflict: control over the work situation. Kornhauser argues that professional scientists seek to maximise their control in terms of 'how', 'where' and 'when' to tackle a project. Industrial management, however, prefer organised research teams working against a time-schedule, and wish to determine at any time which project shall be given priority. Thirdly, Kornhauser suggests there are conflicts over incentives. Professional scientists prefer rewards related to their professional needs—more autonomy, 'free-time',

6

equipment, freedom to attend scientific conferences, etc.—rather than traditional organisational rewards such as promotion to management, albeit research management. Lastly, there is a conflict centred around the responsibility for the utilisation of the scientists' 'products'. The professional feels some ethical responsibility for his knowledge, and the use to which it is put; management considers this is their domain, and that the decision should be mainly determined by commercial considerations.

CONFLICT OR ACCOMMODATION?

Two main lines of solution have been suggested. Some writers have argued that the initial conflict can be at least moderated by modifications in the attitudes of the professional scientist; by a measure of re-socialisation, or by changes in the education of undergraduates. Abrahamson[21] suggests that the initial conflict between professionals and administrators centres around the different expectations each has about the appropriate level of autonomy a researcher should be granted. The new entrant, fresh from graduate school or university typically wants too much autonomy; research administrators are typically prepared to give too little. However, over time, research management becomes prepared to grant more research freedom to the scientist, who indicates his ability to be responsible, in management terms, without tight supervision. In addition, the scientist usually comes to desire less research freedom than he did previously, because he comes to realise that in an industrial setting, his earlier aspirations were not realistic. Thus, by this gradual mutual shift in position, a satisfactory accommodation between professional and employer is forged.

Whyte,[22] in his polemic against the anti-individualism of the modern corporation, sees the process rather as one of the deliberate company indoctrination of 'eccentric' scientists, through an emphasis on team planning, group research, harmonious inter-personal relations, and insistence on company loyalty. Marcson,[23] described the practice of a large electronics company which sent scouts to universities a full year before graduates' studies were programmed to be completed. These scouts attempted to select scientists with interests in projects that higher management had already decided to have investigated. Once employed, the scientist's research aspirations were continuously redefined by supervisors, using the criteria of what was technically and personally feasible, and what was financially supportable. Soon the recruit, through this 'cooling-out'[24] process, comes to accept the definition of the situation as established by higher management. This process of 'acculturalisation' thus reduces some of the strain of the new recruit with research interests not completely fitted to the requirements of the company. It was such company strategies that Whyte referred to as 'indoctrination'.

The second main line of argument has been to suggest that the distinctive needs of the scientist should be recognised, and various means adopted to accommodate him into industrial organisation. For example, Shepard[25] has argued that those companies which employ numbers of scientists in more basic and applied research, should develop a dual system of status hierarchies, with loose formal and informal links between the research sections and higher management via research directors and group leaders. In this way, scientific control over research work could be increased. However, the evidence from America is that this strategy is not widely pursued, and even where it is, higher research management are often selected simply because they reflect the firm's definition of research.[26] Laporte[27] found in his study of an aerospace industrial complex, that accommodation between scientists and managers was arrived at by a process of structural separation of scientists, managers and administrators. Thus, the research laboratories were able to establish some degree of functional autonomy.[28] This led to a reduction of accounting and administrative routines that were only suitable for production and not research. By such structural rearrangements, the research staff became more integrated.

Kaplan,[29] in a study of scientific productivity, argued that laboratories will tend to be more effective the more they satisfy five requisite conditions. Firstly, by showing enthusiasm for new ideas and by being receptive to innovation. Secondly, be removing the typically high level of pressure on employees that characterises other types of production. Thirdly, by tolerating 'odd-balls', researchers who do not fit in with the strictness of organisational conformity. Fourthly, by giving researchers more freedom to choose problems and by allowing them to change research direction. Lastly, by devising suitable professional incentives and rewards, such as more freedom to attend scientific meetings, both nationally, and internationally.[30]

Kornhauser summarises the position by suggesting that both the need for 'structural autonomy', which, he argues, characterises the professional groups of scientists, and the need for 'functional integration' which the overall organisation requires, can only be resolved by mutual accommodations. Each should recognise their mutual interdependence—the scientists rely on the organisation for resources; the organisation relies on scientists for innovations. This reciprocity should be the foundation on which better understanding is built. Thus he suggests industry should allow more basic research in return for more developmental effort; research concerns should be controlled by colleagues although hierarchically co-ordinated; industry should provide scientific career ladders as an option to the normal administrative career ladder. Whilst these accommodations tend to favour scientists, the latter should not press for an extension of responsibilities over the utilisation of their intellectual products, since this would not allow for an establishment of some measure of 'functional autonomy'. But, unless they establish a degree of structural separation from production, they will be unable to untangle themselves, and their research, from the commercial-nexus.

SCIENTISTS, PROFESSIONALS OR ORGANISATION MEN?

Now a major weakness in the argument so far is the assumption that all men with a BSc or PhD are 'scientists', in the sense that they are dedicated to the pursuit of scientific knowledge, and need an environment which allows their creativity to blossom. Both the impressions of those who have anything to do with scientists, and a growing volume of research, concur in supporting the view that this is too simple a picture.[31] The case rests in part on a methodological weakness. We cannot infer motives and values from behaviour. The fact that an individual acts the role of a scientist does not by itself tell us what are his values and motivations.[32] Yet it is this kind of assumption which underlies much of the argument about the bureaucratisation of science, and the conflicts between science and industry.

If then, as both experience and the literature suggest, the term 'scientist' embraces a rather heterogeneous group of individuals, only some of whom are dedicated to the pursuit of knowledge, what are their characteristics—their needs, values, motivations, aspirations? What, in short, are the meanings which science can have for different individuals? What leads some to careers in industry? And what are the specific rewards, strains, frustrations, for different kinds of scientist in the range of industrial roles, from fairly fundamental basic research to quality control, trouble-shooting, or technical roles?

In short, to penetrate the complexity of such issues, we need a clearer picture of the nature both of science and of scientists, and of the roles which scientists play. Only then can we explore compatibilities, strains, conflicts, and the wider implications of the growth of industrial science. And it is to such issues that we turn in the remainder of this study.

THE RESEARCH PROJECT

In order to obtain empirical evidence on such problems, two main surveys were carried out.[33] Firstly, we interviewed a number of research scientists and administrators in nine industrial research laboratories. Our criterion for selection was simply that they were qualified chemists, and employed in an R and D department. We chose chemists because they constituted the largest single group employed in industry and we kept to one discipline because we wanted to control for as many variables as possible. Our findings relate therefore, specifically to graduates in chemistry. In addition to the interviews, we contacted a much larger number by postal questionnaire and followed these up by a further series of intensive interviews.[34]

The second main project[35] reported here was a postal questionnaire survey of undergraduate and postgraduate students in chemistry at three universities. The object of this part of the inquiry was to try to penetrate more deeply into an understanding of the factors which led individuals to choose to study science, why some become more committed to science than others, and why some choose a career in industry and others in the universities. This part of the research is reported mainly in Chapters 3 and 4.

NOTES

1. D. J. de Solla Price, *Little Science—Big Science* (1963). For figures relating specifically to Britain, see T. R. Pike, *History of Scientists in Britain*, unpublished M.Sc. Econ. thesis, Univ. of London (1961).
2. On these issues, see D. J. de Solla Price, op. cit. (1963); N. W. Storer, *The Social System of Science* (1966); W. O. Hagstrom, *The Scientific Community* (1965).
3. *The Flow into Employment of Scientists, Engineers & Technologists* (Swann Report), Cmd. 3760 (1968), p. 46.
4. ibid., para. 44.
5. ibid., para. 55.
6. See also, S. Hatch and E. Rudd, *Graduate Study and After* (1969); R. K. Kelsall, *A National Survey of University Graduates qualifying in 1960*.
7. ibid., p. 89, Table 14.
8. See, for example, R. Blauner, *Alienation and Freedom* (1964).
9. A number of works, based on Maslow's 'self-actualisation' motivational model believe individuals are prevented from self-realisation in modern organisations. The implicit thesis in these writings is that the individual seeks to mature, or achieve actualisation at work, but is prevented by organisational demands. The individual is thus less than his 'essence'. See C. Argyris, *Integrating the Individual and the Organisation* (1964); F. Herzberg, *Work and the Nature of Man* (1966); D. McGregor, *The Human Side of Enterprise* (1960); A. Kornhauser, *Mental Health of the Industrial Worker* (1965). These works have been coming under criticism recently; see particularly G. Straus, 'The Personality-versus-Organisation Theory,' in Sayles, L. R. (ed.) *Individualism and Big Business* (1963), pp. 67–80. Other writers have tended to stress the conformity, subservience, and apathy of organisational employment, and its implicit threat to liberal-democratic values. See R. Presthus, *The Organisational Society* (1962); W. H. Whyte, *Organisation Man* (1957).
   It is also interesting to note that of the two historical themes open to modern writers on bureaucracy, Weber's 'pessimism' should have been wholly stressed, whilst Saint Simon's 'optimism,' his belief that organisations were a liberating force, helping to lift the yoke of work from man's back, has been ignored. See on this, A. W. Gouldner, 'Metaphysical Pathos and the Theory of Bureaucracy,' in A. Etzioni, (ed.) *Complex Organisations* (1962), and 'Organisational Analysis', in R. K. Merton *et al.*, *Sociology Today* (1959), pp. 400–28.

10. W. Kornhauser, *The Politics of Mass Society* (1960).
11. The term is being used here in the sense defined by R. Blauner, op. cit. (1964).
12. See P. M. Blau and W. R. Scott, *Formal Organizations* (1963).
13. Sociology is constantly faced with this problem. Terms such as bureaucracy, profession, working-class, are part of the public language of society. Attempts by sociologists to communicate with a wider audience constantly face the difficulty by specifying a more rigorous definition. Alas, the tasks is not made easier by the too frequent lack of consensus among sociologists.
14. We shall later raise objections to this.
15. P. F. Drucker, 'Management and the Professional Employee,' *Harvard Business Review*, XXX (1952), pp. 84–90.
16. Since that date, Drucker has revised many of his ideas, see 'Twelve Fables of Research Management', *Harvard Business Review*, XLI (1963), pp. 103–14.
17. D. G. Moore and R. Renck, 'The Professional Employee in Industry', *Journal of Business* (Jan., 1955). A similar explanation for low morale among scientists can be found in Opinion Research Corporation, *The Conflict between the Scientific Mind and the Management Mind*, Princeton Opinion Research Corporation (1959).
18. T. Burns and G. M. Stalker, *The Management of Innovation* (1961); P. Brown, 'Factionalism and Organisational Change in a Research Laboratory,' *Social Problems*, III (1956), pp. 235–43; C. Shepard and P. Brown, 'Status, Prestige and Esteem in a Research Organisation,' *Administrative Science Quarterly*, I (1956), pp. 340–60.
19. C. D. Orth, 'The Optimum Climate for Industrial Research,' *Harvard Business Review* (Mar.–April, 1959), pp. 55–64. See also, N. Kaplan, 'Some Organisation Factors Affecting Creativity', *IRE Transactions on Eng. Management*, VII (1960), pp. 24–9; L. Meltzer, 'Scientific Productivity in Organisational Settings,' *Journal of Social Issues*, XXII (1956), pp. 32–40; E. Raudsepp, *Managing Creative Scientists and Engineers: A New Management Programme for Increasing Professional Creativity in Industry* (1963), esp. Ch. 4.
20. W. Kornhauser, *Scientists in Industry* (1962).
21. M. Abrahamson, 'The Integration of Industrial Scientists,' *A.S.Q.*, IX (1964), pp. 208–18.
22. W. H. Whyte, *Organisation Man* (1956).
23. S. Marcson, *The Scientist in American Industry* (1960).
24. This 'cooling-out' process seems to be employed in many situations and organisations. See for exposition and example, E. Goffman, 'Cooling the Mark Out: Some Aspects of Adaptation to Failure,' *Psychiatry* XV (1952), pp. 451–63; and B. R. Clark, 'The Cooling-Out Function in Higher Education,' *A.J.S.*, LXV (1956), pp. 559–76.
25. For example, see H. A. Shepard, 'The Dual Hierarchy in Research,' *Research Management*, I (1958). See also, W. Kornhauser, op. cit., pp. 117–30.
26. W. Kornhauser, op. cit., pp. 56–73.

27. R. T. Laporte, 'Conditions of Strain and Accommodation in Industrial Research Organisations,' *A.S.Q.*, X (1965), pp. 21–38.
28. Briefly, this concept refers to a relationship between a sub-system and the major system in which it forms part, such that the latter is able to obtain gratification for some of its needs outside the relationship, whilst the system is more dependent upon the sub-system to meet its needs. See A. W. Gouldner, 'Reciprocity and Autonomy in Functional Theory,' in L. Gross, *Symposium on Sociological Theory* (1959), pp. 241–70.
29. N. Kaplan, op. cit. (1960).
30. Examples could be multiplied. E.g., Baumgartel found a high association between participatory styles of supervision and high motivation to work. See Chapter 7 for a more detailed discussion.
31. See Chapter 2 for a review of the literature.
32. As we shall argue in more detail in the next chapter, the fact that a social system such as *science* has institutionalised certain values, does not permit us to assume that all individual *scientists* have internalised these values. This distinction was, of course, pointed out by Parsons. See also: G. C. Homans, 'Bringing the Men Back In,' *A.S.R.*, XXIX (1964), pp. 809–18; D. Wrong, 'The Over-Socialised View of Man,' *A.S.R.*, XXVI (1961), pp. 184–93; R. Bendix and B. Berger, 'Images of Society and Problems of Concept Formation in Sociology,' in L. Gross, op. cit., pp. 92–120.
33. The field work was conducted between 1964 and 1967.
34. See Appendix 1 for further details of this inquiry.
35. See Appendix 3. This part of the inquiry was conducted by Dr Box and its findings presented as a PhD thesis, Univ. of Lond. (1968).

Chapter 2

Science and Scientists

In the last chapter, it was suggested that there is no necessary connection between being qualified and employed as a scientist, and a full identification with the norms of science. That is to say, not all who work as scientists are necessarily committed to the advancement of science. Now this is a key issue, since, as we have seen, the whole question of the conflicts and tensions experienced by scientists in industrial employment turns on the extent to which those employed in industry subscribe to the values and norms of science. It is 'dedicated' scientists whom we would expect to experience most strains. Our first task then will be to attempt to explore more fully the characteristics of science and the nature of the scientific role. Only then, when we have a clearer understanding of the nature of science and scientists, will we be in a position to explore empirically the articulation between science and industry in the specific context of industrial research laboratories.

SCIENCE AS A SOCIAL SYSTEM

It is interesting to observe that studies of the nature of science have typically been written by non-scientists. In particular it is philosophers, rather than scientists, who have been preoccupied with questions about the logic and methodology of science. And it is such philosophical perspectives which have coloured prevailing conceptions about the nature of science, emphasising the 'scientific method' as its characteristic feature. But such studies have painted only one side of the picture. They have directed our attention away from the fact that science is above all a social activity. Scientists typically communicate their findings to other members of the 'scientific community', particularly to other members of the numerous 'invisible

14

colleges' of scholars interested in a particular field. Indeed, this emphasis on the imperative to publish, to share findings, is one of the most distinctive features of science.

It is this aspect of science which has attracted a good deal of attention from sociologists and historians[1] in recent years. But it leaves many problems unsolved. All are agreed on the crucial role of publication. Science is above all 'public' knowledge. 'The objective of science is not just to acquire information . . . ; its goal is a *consensus* of rational opinion over the widest possible field.'[2] And the complex machinery of journals and referees[3] is the method by which the work of the individual scientist receives the imprimature of the scientific community, and his individual contribution is recognised *as* a contribution.[4]

The goal of science then is 'public knowledge'. And it is the 'scientific community' which pursues this goal. Now the interaction of numbers of individuals in the pursuit of a common goal requires a set of rules to regulate their activities and interactions. Indeed, one of the first attempts (Merton, 1937) to look at science as an organised social system sought to identify the major norms of science. The first of these, the norm of *communality*, underlines the imperative to publish, directs scientists to regard their discoveries as something they should share with others and not as their own private property. Secrecy would fatally hinder the consensual process.[5]

Given the institutionalised goal of science then, the norms follow as necessary rules for the effective pursuit of the goals. If the goal is consensus on public knowledge, then each scientist must look critically at the contributions of his colleagues. Only those contributions which can withstand the scrutiny which follows from such *organised scepticism* become part of the consensus. Further, the criterion for assessing a contribution must be confined to those relevant for science, and must exclude, for example, the nationality of the scientist, or the geographical location of the work. It is this norm of *universalism* together with the norm of *communality* which gives science its international character.

Finally, the norm of *disinterestedness* exhorts the scientist to seek only the rewards of science—the recognition which follows from the acceptance of his contribution. Indeed, as

Hagstrom argues, we may usefully think of the social system of science as a system in which free gifts[6] of knowledge to the scientific community are exchanged for recognition. Certainly, such rewards form an important element in the social organisation of science—in the form of eponymy, fellowships, and a variety of honours. Moreover, it is discovery which is rewarded —hence the anxieties and the often bitter disputes over priorities.[7]

The norm of disinterestedness is particularly important in that it limits the rewards for which the scientist may legitimately strive, and ensures that the research which he undertakes is determined by scientific criteria. He is more likely, for example, to be rewarded by recognition for work which is of interest to other scientists, rather than work of importance primarily to industry. Hence the condemnation of the search for popularity.[8] In short, the value of a contribution to *knowledge* can only be determined by other scholars, not by its market value, or by the conferment of political honours.

To describe the goals of a social system, and the rules which regulate the conduct of its members, raises a series of problems about behaviour at the level of the individual. Individuals can play the game according to the rules because (a) they may be anxious to avoid negative sanctions, (b) they think the rules are 'right' or morally binding, (c) they calculate that it will be rewarding in the sense that felt needs will be satisfied.[9] Why do scientists comply with the norms? Perhaps because the lengthy socialisation at school and university produces persons who become strongly committed to the values and norms of science. As we saw in Chapter 1, this is the assumption underlying much of the recent work—that the science graduate has accepted the values and norms of science. As Hagstrom argues, it is certainly a highly selective process, so that those who survive to complete their PhDs are likely to be strongly identified with science. Alternatively, Merton argues that the norms are accepted as binding simply because they are seen to be functionally necessary. Subsequent writers, notably Storer and Hagstrom, have explored this aspect of the problem more fully and offered alternative explanations. Hagstrom argues that breaches of the norms will lead to the withholding or withdrawal of recognition for the gifts of knowledge. In particular, publication in journals which pay contributors will not earn recognition. A somewhat similar conclusion is reached by Storer, from a slightly different starting point. He goes somewhat further back than Hagstrom, and starts from the hypothesis that what characterises the scientist above all else is his desire to be creative. But confirmation of his creativity can only be achieved through recognition by the scientific community. Only his peers are competent to judge the worth of his contribution. And only by conformity to the norms can the individual scientist earn the recognition of his peers. For Storer then, the norms of science are seen to be relevant to the scientist's search for a competent response, rather than for the advancement of science as such.[10]

But such explanations can hardly account for other features which characterise scientists, in particular the often intense enthusiasm and excitement—the 'emotions' of science—nor the concern for priorities which seem to have antedated the emergence of the scientific community and the institutionalisation of the norms of science.[11] Eiduson gives many examples of the intensity of these emotions of science. 'I think the biggest thrills I've had in my life, the most satisfying things I've ever done, are the few little discoveries I've made . . . once in a while you make this decisive step that you know is going to change the future of science, and I think it's a kind of thrill that nothing else can equal. . . . If a bright student can experience this, you can't stop him; if he experiences it once, he wants to do it again.'[12] Such sentiments echo the words of Max Weber in his essay on Science as a Vocation. 'Without this strange intoxication, ridiculed by every outsider; without this passion . . . you have no calling for science and should do something else.'[13] To understand such characteristics we must go beyond a study of the social system of science and explore the factors in the making of a scientist.[14]

Finally, we must consider the possibility that conformity to the norms and values of science may be nothing more than the recognition that these are the rules of the game. And if one has chosen science as a career, because, for example, life in a university is seen as an attractive way of earning a living, then one must keep the rules.

ROLES AND IDENTITIES

Put very simply, what we are saying is that it is not possible to make valid deductions about an actor simply from the fact that he is playing a particular role. Behavioural conformity to the norms of science is one thing. Accepting them, internalising them as morally binding, is another. The conceptual tools for this kind of distinction have been considerably extended in recent years. The notion of *role* refers to the set of behavioural expectations attached to a particular social position. The concept of *identity*[15] enables us to take account of the characteristics of the actor which he brings with him to his role performance. Identity refers to what a man *is*, to how he sees himself; role refers to how he is expected to behave. We need to investigate directly the motives and values of the individual.

But the situation is more complex than this. As actors in social situations, we each of us occupy a number of roles. We may simultaneously occupy a role in the economic system—as teacher, clerk, scientist—and in the kinship system as husband, father, brother. As teacher, scientist, clerk, we conform more or less with the socially established expectations for that role. Within organisations, roles are spelled out in some detail so that everyone knows with a minimum of ambiguity what is expected of him, and conforms, more or less, to the set of role expectations as he perceives them.[16]

But this dramaturgical analogy can be carried further. Just as the actor on the stage may pour himself completely into some roles, so for the social actor, a particular role may be totally absorbing, one in which he invests his total self, leaving little over for other roles. This is the situation which may characterise the dedicated scientist. He has embraced[17] his role, he *is* a scientist. In fact, of course, such 'total' roles are few. Even the scientist, unlike the monk, is likely to occupy the role of husband and father. There is, that is to say, a part of the self left outside. He has not invested quite all—he *is* a husband—and his other roles involve parts of his total identity. However important (salient) his role as scientist may be as involving a part of his total self, he is nevertheless at the same time, a husband. That is, while acting the role of scientist,

18

he carries with him, to use Gouldner's term, the *latent identity*[18] of husband.

In other words, we need to explore not only the way in which the actor perceives any particular role, but also the relative saliency of the scientific role in relation to other roles. Moreover, even as a scientist, the individual may occupy more than one scientific role, each with a specific cluster of expectations for behaviour. The scientist may also, for example, be an employee. The next step, therefore, is to attempt a specification of the main types of role that the scientist may occupy.

SCIENTIFIC ROLES

Most of the literature on the sociology of science has focused attention on the traditional role of the scientist in the scientific community. But, as we saw in Chapter 1, an increasing proportion of graduates in science are now employed in industry, government research institutes, teaching. As such, it is less likely that they will either think of themselves, or be defined as being members of the scientific community. In fact, they occupy roles in distinct sub-systems of society, each with specific goals and norms. The primary function of the industrial research laboratory is not the advancement of knowledge as such, but rather the application of scientific knowledge and skills to the development of marketable products. This may involve discovering new knowledge. But the objective is not to contribute such knowledge to the current store. On the contrary, the general industrial practice is to control publication, for example, by delaying until covered by patents.[19]

How then, can we differentiate the role of the industrial scientist from that of the traditional scientist? The crucial distinction clearly centres on the difference in the use of knowledge. The scientist's contributions to knowledge are evaluated by different individuals, using different criteria. The criterion in industry is usefulness to the firms products, and the audience is his colleagues and managers in the firm, and not the scientific community. The scientist is functioning in the economic sub-system. The goals are marketable products, and the criteria and rewards are economic. This does not mean that the norms of science are completely irrelevant, but they are likely to be less

19

important. Organised scepticism is crucial, where the only competent response is recognition for contributions to knowledge. But where the test is a pragmatic one—that the process works—this becomes the more important criterion. This difference in role expectations is brought out very clearly by the following remark by one of the scientists we interviewed:

'I and my colleagues are often dumbfounded at the way in which half-baked ideas, pushed into production on half-suitable machinery, nevertheless make a mint of money. I think that none of us fail to appreciate the bed-rock quality of pure science, but we also have learned that a saleable imperfect product keeps body and soul together with more force than does a product that is in all respects too exotic.'

The norm of communality, however, is clearly inappropriate. The main object of industrial research is the development of marketable products. And publication is expected to take second place to the economic interests of the firm; as one of our respondents clearly recognised:

'Except on the rather rare occasion where pure research occurs in industry, it is unreasonable to expect firms to encourage publication when their competitors may benefit thereby. I would like to think otherwise, but it does not appear realistic to do so.'

TABLE 2.1
*Main type of research activity of
sample of industrial chemists*

	Per cent
Basic research	5
Applied research	28
Development	39
Service	21
Other	7
	100
	($n = 403$)

In practice, of course, the roles of industrial scientists are more highly differentiated than this simple picture implies. At one end of the R and D spectrum, we may have scientists engaged in fairly long-term fundamental work, with no immediate pay-off in prospect. Such work is more prevalent in those industries such as the pharmaceutical industry where the market position of the firm depends on product innovations which rest heavily on advances in knowledge. At the other end of the research spectrum, there is the servicing of established production processes, including quality control and trouble shooting, where scientific knowledge is being applied through standard procedures. (See Table 2.1 for the distribution of our sample of chemists between the major R and D functions.)

One final type of role which the industrial scientist may be called upon to play is a strictly non-scientific role, but where a knowledge of science may be of value. Thus, in production management and technical sales or purchasing, a knowledge of science may be essential for effective performance even though the job does not involve carrying out any of the normal functions of a scientist.

We have then three main types of scientific role. The first is essentially an *academic* role. The pursuit of knowledge is the major function of academics. The norms and values of science are shared by other academics in many disciplines. This is not to say that there are no differences. It is only in the natural sciences that the community of scholars is so crucial and that the emphasis is on achieving consensual knowledge. In the arts, the aim is more frequently private experience rather than public knowledge. Moreover, the audience whose response is sought extends beyond a narrow group of experts.[20]

The second main type of scientific role is characterised by the application of scientific knowledge and skills and is exemplified by the industrial scientist in R and D. Although we recognise that new knowledge may be an outcome of such activities, this is not the primary object or expectation. It is solutions to practical problems which is sought, not knowledge as such or for its own sake. It is this which differentiates the *professional* role from that of the academic. This type of role is, of course, familiar to us as that of the professions. It is professionals[21] who are concerned with the application of expertise, in the form of knowledge and skills. And although engineers, doctors and architects draw on a range of knowledge from a variety of sciences, they are not, as professionals, primarily preoccupied with the fundamental work for the advancement of knowledge in, say, the strength of materials, or cell structure. Thus an engineer may function in an academic role, undertaking research into heat-engines, and primarily concerned with the advancement of knowledge. Or, he may function as a professional in the design of an engine.

Finally, there is the third type of scientific role which can be generalised most simply as an *organisational* role. Organisations are typically a means for achieving a high degree of division of labour by allocating specific tasks (roles) and co-ordinating the activities of numbers of individuals for the optimum achievement of organisational goals. The organisational role is essentially that of an employee. The individual carries out his specified tasks[22] in exchange for a salary or wage. The scientist employed in production management or technical sales is carrying out an organisational role, however much discretion he may be allowed and however loose his job specification.

Now clearly, such distinctions are too simple. In particular, the notion of the scientist in R and D as a professional runs into difficulties. As an expert, he is likely to enjoy a measure of autonomy, for only he can decide what to do on questions of professional expertise. But he is also an employee, occupying a role in an organisation. And there are different rules for these roles. The professional demands autonomy. The employee must comply with company directives. Similarly, the scientist undertaking developmental research may well come across new knowledge which he considers publishable. Now, there is no reason why individuals should not act two roles simultaneously. It is possible to play two games at once. But we must play each game by its own rules. For example, we may play a game of tennis and at the same time seek to date our partner. This doesn't entitle us to serve double faults, but we may decide not to try very hard to win the tennis game. Similarly, the rules for the academic knowledge game must be complied with if the researcher is to make a contribution. And this means going beyond what is required in the performance of his industrial roles. It may not be easy to play two games at once. The moves required for one may conflict with those required for another. Just what problems are involved will be explored more fully in Chapters 5 and 6.

SCIENTIFIC IDENTITIES

So far we have spoken only of role requirements, the expectation to publish in the role of the academic scientist, to exercise a high level of expertise for the professional, and to give a fair day's work for a fair day's pay for the employee. What of the characteristics of those who occupy such roles? As we saw in Chapter 1, much previous work in the sociology of science has failed to distinguish between the institutionalised values of science, and the internalisation and acceptance of such values by scientists. It is problematic and by no means self-evident, that all those who possess a BSc have full internalised and accepted the values of science.

Much of the discussion around the contemporary crises in science stems from an assumption that the label 'scientist' identifies individuals who have, in fact, made such values their own. It is in such situations that the conflict between the values of science and those of industry are indeed likely to occur. But we must also consider the possibility that, just as there is a variety of scientific roles, so there is more than one scientific identity. Furthermore, if it could be demonstrated that there is a reasonable matching between the identities and the roles available, then the alleged strains inherent in the increasing deployment of scientists in industry might be minimised.[23]

Just as we have identified three main types of scientific role, so we can construct a typology of scientific identities corresponding to such roles. These will, of course, be ideal-type constructs, but we will hope to demonstrate that such a typology can be supported empirically. We are suggesting, in short, that there is a type of scientist whose characteristics are broadly congruent with each of the main types of role, as set out schematically in Figure 2.1. A starting point for any such typology is the frequently observed distinction between professional and other orientations. For example, Marvick[24] made a distinction between the 'specialist' and the 'institutional'

Scientific roles and types of scientist

Roles	Academic (knowledge)	Professional (application)	Non-scientific
Identities	Public	Private	Organisational

FIGURE 2.1

orientation. The former 'is one in which professional expertise is given primacy. Stress is placed on furthering a career in the profession, avoiding executive posts, and not being overly concerned whether professional employment is found in public or private organisations. Preoccupied with matters of skill gratification, this type tends to be indifferent and detached in respect to the material benefits and social life of the work establishment'. The latter orientation is characterised as being 'place-bound; gratifications are sought in the personal benefits available within the work establishment. This type tends to be indifferent to benefits that are derived from the application of skills to a task'.[25]

Another much-employed dichotomy referring essentially to the same phenomena is that coined by Merton: 'cosmopolitan-local'. This distinction was taken over by Gouldner[26] and used to describe types of organisational employees, particularly in the academic world. Whether the two types are called 'professional-organisational', or 'cosmopolitan-locals', or 'science oriented—company oriented', it has been assumed by the writers that there are two independent dimensions of orientation —the institution (profession, science) and the organisation. The clearest statement of the alleged relationship between the two orientations is to be found in Caplow and McGee:[27] 'Today, a scholar's orientation to his institution is apt to disorient him to his discipline and to affect his professional prestige unfavourably. Conversely, an orientation to his discipline will disorient him to his institution, which he will regard as a temporary shelter where he can pursue his career as a member of the discipline.'

More recent studies have indicated the possibility of rather complex relations between individual orientations and organisations. Glaser's study of a medical research laboratory showed that, where organisational goals are the same as the goals of science, then it was possible to have a type of scientist who was both strongly oriented towards the institution of science and at the same time strongly oriented towards his employing organisation.[28] Blau and Scott,[29] agreeing with Glaser, also argue that 'localism' and 'cosmopolitanism' are not necessarily distinct. They suggest, 'a commitment to professional skills will be associated with low organisational loyalty *only* if professional opportunities are more limited in the organisation under consideration than in others with which it competes for manpower . . . only if it is the structural conditions of the organisation rather than the structure of the profession that restricts opportunities for professional advancement do we expect professional commitment to be accompanied by a cosmopolitan orientation.' A similar point was made by Avery.[30] He argued that, ' . . . the career question confronting the technical man is not, typically whether to commit himself wholly to localism or cosmopolitanism. Rather he is likely to be constrained to try to extract advantages from both sources.'

However, despite the body of sociological literature concerning different types of scientists, such typologies have not so far been systematically related to the norms of science outlined by Merton. It is our intention to attempt to construct such a typology and to test its validity empirically. It can be seen from Figure 2.2 that each of the norms of science can be described in terms of its meaning for the individual scientist.

Norms of science and the individual counterparts

Communism	:	Attaches importance to publication
Organised scepticism	:	Attaches importance to autonomy at work
Disinterestedness	:	Is concerned only with a career in science
Universalism	:	Is a supra-nationalist

FIGURE 2.2

If all four variables are dichotomised into 'high' and 'low' we would have sixteen types of scientists. However, there are reasons for thinking that in fact, only a more limited number of combinations will occur. We suggest that the norm of 'communism' is the most important in the social system of science, since it is central to the gift-exchange-recognition process. Whereas science could exist where the other norms are weak, it could not exist unless each individual believed communication of his work to his colleagues to be of key importance. Because of the centrality of this norm, we would expect it to have an umbrella effect, that is to say, if a scientist attached importance to publication, then we would expect that he would also attach high importance to the other norms of science. We shall refer to this type of scientist as the *public* scientist.

We may distinguish scientists from many other professionals, such as architects, by this key criterion—that they seek recognition from their peers through publication. But confirmation of the worth of a contribution can come from one's immediate colleagues, or, as with the engineer, from the fact that the product works. Such scientists, who while they attach importance to disinterestedness and organised scepticism do not seek recognition and confirmation from the scientific community, we have called *private* scientists. It is these who correspond to other professionals.

In contrast to these, there is a third major type who pursues scientific research as a means of earning a comfortable living. And although he may find his scientific work in R and D interesting and rewarding, he is not committed to pursuing a career in science, nor does he seek recognition through publication. It is such men, we shall argue, who are willing to move out of research into some other organisational role, such as management, and whom we shall call *organisational* scientists.[31]

In suggesting these three types of scientists (Figure 2.3) we are also suggesting that certain patterns would seem very unlikely. For instance, it would be difficult to conceive of a scientist who placed considerable importance on publishing his work, and yet placed little value on what other scientists think of it, had no interest in having freedom at work, and whose reason for publishing was to obtain rewards from some other institutional sphere.

FURTHER CLARIFICATION OF THE TYPES OF SCIENTISTS

The public scientist is one who is dedicated to developing his skills as a scientist and to making a public contribution to a body of knowledge. When he prepares a paper for publication, and offers it for that purpose, he is not only claiming to advance

A typology of scientists

Type of scientist	Importance attached to norms of science			
	Communism	Disinterestedness	Organised scepticism	Universalism
Public	+	+	+	+
Private	−	+	+	−
Organisational	−	−	±	−

+ indicated attaches importance to norm
− indicated attaches little importance to norm

Where importance is attached to one norm, it is hypothesised that importance will be attached to norms to the left of it.

FIGURE 2.3

knowledge, but he is claiming to be a particular kind of scientist. His definition of himself as this particular kind of scientist can only be maintained if it is continually being confirmed by a wide audience of scientific colleagues. This kind of scientist sees himself caught up in the historical development of a subject, with his contribution making a further addition. But, in order to have his contribution admitted, he has to make his claim widely known, mainly in academic journals. In this way, his credentials are recorded and scrutinised by significant others, who in turn, if they are satisfied, confirm his claim to be one of them. The public scientist is therefore— using Parsonian language—characterised by his 'collectivity-orientation'; he desires to share his knowledge with, and be part of, a corporate body of fellow scientists, the scientific community.

In addition to this 'collectivity-orientation' the public scientist is also characterised by a disinterestedness in other rewards, which takes the form of a deep 'commitment' to science. This commitment has two dimensions: socio-psychological, and spatial-temporal.

Goffman[32] has argued that an actor's relation to his role can be measured along two dimensions. Firstly, there is the extent to which he feels attracted to and attached to the role, and the extent to which therefore he identified himself with the role. This he terms *attachment*. Secondly, there is the extent to which escape from the role is possible. In some roles, the cost of vacating may be very high, in personal or material terms. This dimension he refers to as *commitment*. Where an actor is both attached and committed to a role, Goffman suggests he can be described as *embracing* it. Now it is this which characterises the relationship of a public scientist to his roles as public scientist. He *is* a scientist; and his allegiance is first and foremost to science—an allegiance above class, race, nationality—and certainly higher than his allegiance to any organisation. It is this characteristic to which Stein refers when he describes such scientists as 'being devoted to their goals and making inordinate sacrifices in order to achieve them; they integrate complex situations into simplified and meaning-ful new developments; and they are dynamic in that they strive for distant goals.'[33]

The spatial-temporal dimension of the public scientist's commitment takes the form of its interpenetration into other life-areas, and its continuity through time. In her study of eminent scientists, Eiduson says that 'internal compulsiveness has dictated the way (they) work at research'. In this investi-gation, almost every man spoke of his ' . . . long hours, his seven-day week, and his complete absorption in his research problems. In fact, these kinds of efforts have become so mechanically affiliated with scientific work, that they seem to have become merged with the values of scholarship, rigour and discipline, and represent the only proper way they can be seriously pursued.'[34] In this sense of commitment, the public scientist is hardly ever away from work; he is constantly, whether in his formal place of work, or not, orientating to objects and others from the perspective of his scientific identity.

It is interesting that this characteristic of the public scientist is one which constitutes a salient part of the image students have of scientists, and which manifests itself in the humorous image of the scientist as being 'the absent-minded professor'.[35]

The privatised scientist, like the public type, is also deeply commited to science as an activity, and equally desirous of autonomy in his work place. However, he does not share the communistic ideal of the public scientist in that he does not place much importance on publication. He is less likely, therefore, to have a strong 'international' orientation. The privatised scientist's commitment is centred around the intense job-satisfaction he derives from tackling and solving scientific problems. Given this commitment, he also seeks autonomy because this is often the only necessary guarantee that he can secure a situation in which this kind of satisfaction might be obtained. He faces the danger that he may find himself supervised by persons who treat him not as an autonomous expert, but as a subordinate employee, in which his expertise will be discounted. In this situation, the scientist will find his work tightly controlled, and this control will not necessarily be compatible with his functioning as a scientist. Since for the privatised scientist it is research curiosity and the hope of research success which constitute his strongest motives, the absence of autonomy will be seen by him as both a threat to his professional expertise, and to his scientific creativity.

Quite apart from any theoretical justification there may be for hypothesising the existence of the privatised scientist, there are several historical developments in science which reinforce the arguments for his existence. These refer specifically to the changing role of publication. There was a stage in the development of science when it was possible to make a personal reputation without publishing much of one's work. But this was only typical while the scientific community remained relatively small and was solidified by interlocking personal acquaintanceships. For in these conditions, it was possible to make a reputation through face-to-face, informal communication. However, as the scientific community grew, this mode of establishing consensus became increasingly impractical, and it was replaced by the scientific paper.[36] The relation between publications and recognition is extremely well documentated,[37] and it would not be unreasonable to suggest that this mode of establishing recognition for contributions to the scientific consensus has been the dominant mode in most industrial countries for the last hundred or more years. But, this is not to say that no scientists can make a reputation without publishing. In sociology for instance, Mead and Simmel are examples of scholars who achieved renown although little of their work was published during their lifetimes.[38] However, it is claimed here that the establishment of a scientific reputation through publication is, or at least has been up until very recently, the modal technique. But the continued rapid development in science and the growth in the numbers of scientists have created a situation in which the system of recognition through publication is no longer so viable. This is because the number of papers published has now become almost unmanageable, and at the same time, the increased specialisation which has accompanied this growth in the scientific population, has had the effect of reducing the audience to whom one's work is meaningful and comprehensible. Hagstrom,[39] has even gone so far as to suggest that the system has already reached a state that can be described as 'anomic'. By this he means that more and more scientists are becoming aware of the 'general absence of opportunities to achieve recognition' through publication.

Two important conclusions emerge from these considerations. Firstly, Hagstrom suggests that many scientists in this situation of anomy 'withdraw' from the scientific community, and emphasise some other activity, such as development or service work in industrial laboratories, or teaching, or administration. It is inferred from this, that many science students will, even whilst they are still students, become aware of these 'anomic' conditions in science and will, as a consequence, have already 'withdrawn' from the arena of competition in which the outcome is determined in terms of publications. Further, many who enter the employment market who have not braced themselves for the conditions which obtain there, may well discover their ambitions centred on recognition through publication to be unattainable. The resulting 'retreatism' or 'withdrawal' is, we feel, conveyed in the label of privatisation'.

Secondly, it has been suggested by Price[40] that the continued growth of science will create a situation in which reputations based upon publications will be gradually replaced by a re-emergence of informal reputations built up from personal interactions. But these 'invisible colleges' as Price calls them—will consist of the cosmopolitans of the scientific community, and not the locals; that is, they will not consist of those scientists who are more or less permanently based in one organisation which lacks a high scientific reputation. As these 'invisible colleges' develop, those other scientists who are not members but who are still interested in science as a personal activity, will become an important group, at least numerically. It can be argued then that those scientists (or students) who are intrinsically involved in science, but who are, for various reasons, unable to establish themselves as public scientists will become 'privatised'. Both the public and private types of scientist are intrinsically involved in their work. The differences between the public and private types are: (i) the public scientist will not be prepared to work in an organisation in which publication freedom is denied; and (ii) the public scientist will make a reputation (or will attempt to make one) through his own publications. The private scientist may, or may not make a reputation among the wider community of scientists. If he does, it will not be by publications, but by the spread of his work through the efforts of disciples or students, or through informal channels of communication. In general, more public scientists would be expected to have a standing in the scientific community than would private scientists. In short, the *professional* role of the private scientist demands that procedures and results 'are to be shared *only* with certain selected individuals whose numbers may vary from none to many, *but never outside the company*'.[41]

For the above reasons then, it seems reasonable to dichotomise the *intrinsically* involved types of scientist into those who place importance on publication and those who do not. The concept of 'private' is employed because it seems to give expression to the idea of withdrawal, retreat and isolation. This was partly suggested by the work of Goldthorpe and Lockwood[42] who, in their study of industrial workers, referred to the 'privatised' type. By this, they meant the worker who was still normatively working class, but who was no longer an integrated member of that class because of his geographical isolation. This seems similar to our private scientist, who, although he still subscribes to some of the norms of science, is no longer actively participating directly in the wider scientific community through public exchanges.

The last major conceptual type of scientist, the 'organisational' scientist, is not committed to a career in science, and neither does he consider that the norms of communism and universalism are important to him. The 'organisational' scientist fastens his aspirations on other occupational positions which he considers hold more rewards and satisfactions than those in science. He has, however, defined scientific activity as the most practical means at his disposal for obtaining one of those desired positions. He is essentially an employee, seeking a rewarding career, but more likely to stress the extrinsic rewards of pay and status. Whether the 'organisational' scientist pursues or desires autonomy at work depends upon it being defined by him as being relevant to his career progress. Because his involvement in science is purely calculative he may define autonomy as important if he considers that the advantages of having autonomy will help him advance to an occupational position outside research.

EMPIRICAL VERIFICATION OF TYPOLOGY

In order to test this typology, we investigated a sample[43] of final under-graduate and post-graduate (PhD) students taking chemistry, and a further sample of graduate chemists employed in industrial research laboratories. We sought to measure their commitments to the values of science by a series of questions, probing the importance which they attached to publication, to autonomy, and their degree of commitment to a career in science.[44]

Among the student sample, 84 per cent fell into the three hypothesised type. For the industrial sample, the corresponding figure was 76 per cent (Table 2.2). The deviant cases are indeed in some cases somewhat puzzling. But it must be remembered that there is an arbitrary element in the dichotomisation into high and low which may account for some of the discrepancy. Moreover, questionnaires are of necessity, somewhat crude measuring instruments and it is possible that

a different set of questions might have proved more discrimina-
tory. However, by far the largest cells are the three predictable
on theoretical grounds. The main discrepancy is the group of
forty-three students (9 per cent) and thirty-six industrial
chemists (10 per cent) who are high on commitment to a career
in science, yet low on publication and autonomy.[45] But it
could well be argued that the question on autonomy was too
rigorous (Question 21, see Appendix 2).

TABLE 2.2

*Types of scientists: importance attached to publication,
autonomy, and a career in science
dichotomised into high (H) and low (L)*

	Publication	Commitment	Autonomy	$n = 501$	Per cent
Students					
Public	H	H	H	212	43*
	H	H	L	15	3
	H	L	H	18	4
	H	L	L	0	—
Private	L	H	H	90	18*
	L	H	L	43	9
Organisational	{L	L	H	36}	24*
	{L	L	L	87}	

				$n = 372$	Per cent
Industrial chemists					
Public	H	H	H	64	17*
	H	H	L	19	5
	H	L	H	34	9
	H	L	L	0	—
Private	L	H	H	89	24*
	L	H	L	36	10
Organisational	{L	L	H	92}	35*
	{L	L	L	38}	

INTERACTION BETWEEN ROLES AND IDENTITIES

It was argued earlier that the characteristics of an actor cannot be inferred from the roles that he plays. It is perfectly possible,[46] for example, for a graduate to choose an academic career for extrinsic reasons—because he values the freedom it confers, rather than because he has become deeply committed to science. Such a scientist will need to publish as a condition for career success, and indeed may well become a successful scientist. Moreover, he will be perfectly aware of the normative expectations of the academic role, and we can expect him to support them verbally and behaviourally. But we would not expect him to be committed to a career in science, and would not be surprised if he left the academic world at some point in his career. Behaviour, that is, is a function of both the actor and his situation. And we need to keep these variables analytically distinct. Thus, in our typology, we seek to distinguish scientists according to the characteristics which they possess, as a prerequisite for understanding differences of behaviour in various situations.

One final point. The above discussion suggests that complete congruence between identity and role is not essential. Nevertheless, some identities are likely to be incompatible with some roles. Indeed, it is precisely because so much of the literature indicated serious strains and incompatibilities between public scientists and industrial roles, that we began our investigation. This whole question will, in fact, be explored in Chapter 5. Meanwhile, it would be useful to reiterate the point that the same individual may occupy more than one role. Thus the private scientist undertaking fundamental work in an industrial laboratory is at least partially involved in an academic role. But he is also an employee. Again, the problems involved in such multiple role-playing constitute a topic for more detailed analysis later.

SUMMARY AND DISCUSSION

Much of the earlier literature on the relations between science and industry has underlined the differences between the nature of science and industry. It has been argued here that these are indeed two distinct social systems, with different goals, values, and norms. Moreover, the role of the industrial scientist is analogous to that of the professional rather than that of the academic. But it has been shown that by no means all graduates in science are academics, or *public* scientists, but many possess characteristics which prima facie qualify them for the performance of professional roles in industry, while many again are oriented towards organisational careers.

In developing our typology of scientist, we understandably took as our point of departure those studies which had examined the characteristics of academic science and scientists. Inevitably, this has resulted in a somewhat negative definition of the 'non-academic' scientists. We characterised the 'privatised' type, for example, by reference to their withdrawal from publication. Such an approach is, of course, not only too crude to pick up all the subtleties of orientations towards science, but it neglects the rich variety of positive satisfactions which may reward the industrial scientist, such as the satisfaction from seeing the results of scientific inquiry incorporated into a successful product. Further studies will no doubt develop more complex characterisations and typologies.[47] Nevertheless, as subsequent chapters will show, the somewhat crude categories employed here proved to be of considerable value in exploring the problems of relating science to industry.

How far then, is there a reasonable congruence between the characteristics of these various types of scientist and the different roles which are available for them in industry? We will explore this problem in Chapter 5. But first we need to understand more fully the process by which different individuals develop different characteristics (identities), and the process of selection and self-selection by which they are channelled into different roles. We turn to these issues in the next two chapters.

NOTES

1. See especially W. O. Hagstrom, op. cit. (1965), and D. de Solla Price, op. cit. (1963).
2. J. Ziman, *Public Knowledge* (1968), p. 9 (italics ours).
3. ibid., Ch. 6. See also W. O. Hagstrom, op. cit. (1965) for a dicussion of the central role of communication. It is important not to over-emphasise the extent of consensus which results from the interactions

within the scientific community. A most perceptive study by Kuhn starts from the observation that paradigms acceptable in the past are no longer so; that while the dominance of one paradigm constitutes 'normal' science, anomalies accumulate, give rise to competing paradigms, and the final triumph of one until in turn this current orthodoxy is challenged. See T. S. Kuhn, *The Structure of Scientific Revolutions* (1962).

4. This does not mean, however, that scientists whose claims to discovery have not been accepted at the time they were offered, are forever wrong. Frequently discoveries come before their time, and their authors have to wait for the climate of opinion to change before they, or their disciples, can gain recognition. The example of Mendel, whose work was rediscovered years later is probably a widely known illustration of this. For a brief but instructive analysis of this and other examples, see B. Barber, 'Resistance by Scientists to Scientific Discovery,' *Science*, CXXXIV (1961), pp. 596–602. On the dilemma in science of the genius and the crackpot; see on this, Storer, op. cit., pp. 116–22. See also, R. K. Merton, 'Resistance to the Systematic Study of Multiple Discoveries in Science,' *European Journal of Sociology*, IV (1963), pp. 237–82; 'Singletons and Multiples in Scientific Discovery,' *Proceedings of the American Philosophical Society*, CV (1961), pp. 470–786.

   Furthermore, recognition is not entirely free from the influence of reputation. A paper by an eminent scientist is more likely to receive recognition than one from an unknown contributor. See R. K. Merton, 'The Matthew Effect in Science,' *Science*, CLIX (Jan. 1968), pp. 56–63.

5. J. D. Bernal, *The Social Function of Science* (1939): 'In so far as any inquiry is a secret one, it naturally limits all those engaged in carrying it out from effective contact with their fellow scientists either in other countries or in universities, or even, often enough, in other departments in the same firm.' (p. 107) and see pp. 150–2. See Kornhauser, op. cit., pp. 73–9. See also on this, E. Shils, *The Torment of Secrecy* (1956); P. F. Lazarsfeld and W. Thielens, *The Academic Mind* (1958); W. Hirsch, 'Knowledge, Power and Social Change, the Role of American Scientists,' in C. K. Zollschan and W. Hirsch (eds.), *Explorations in Social Change* (1964); Barber, op. cit., pp. 91–2, 179–80.

6. This is stated in both Storer, op. cit., and Hagstrom, op. cit. (1965), see Chap. 2: 'Whenever strong commitments to values are expected, the rational calculation of punishments and rewards is regarded as an improper basis for making decisions.' p. 21. 'The organisation of science consists of an exchange of social recognition for information. But, as in all gift-giving, the expectation of return gifts (of recognition) cannot be publicly acknowledged as the motive for making the gift. A gift is supposed to be given, not in the expectation of a return, but as an expression of the sentiment of the donor toward the recipient.' (p. 13). See also N. J. Smelser, 'A Comparative View of Exchange Systems,' *Economic Development and Cultural Change*,

VII (1959), pp. 173–82, for a study 'which suggests that the gift mode of exchange is typical not only of science but of all institutions concerned with the maintenance and transmission of common values, such as the family, religion, and communities'; A. W. Gouldner, The Norm of Reciprocity', *A.S.R.*, XXV (1960), pp. 161–78; M. Mauss, *The Gift: Forms and Functions of Exchange in Primitive Societies* (1954), pp. 40 ff., *et passim*; C. S. Belshaw, *Traditional Exchange and Modern Markets* (1965).

7. R. K. Merton, 'Priorities in Scientific Discovery: A Chapter in the Sociology of Science,' *A.S.R.*, XXII (1957), pp. 635–59. Although other writers have discussed the norms of science, all of them have recognised an enormous debt to Merton's initial work. See H. A. Shepard, 'The Values of a University Research Group,' *A.S.R.*, XIX (1954), pp. 456–62; 'Basic Research and the Value System of Pure Science,' *Philosophy of Science*, XXIII (1956), pp. 48–57; F. Reif, 'The Competitive World of the Pure Scientist,' *Science*, CXXXIV (1961), pp. 1957–62; S. S. West, 'The Ideology of Academic Scientists,' *IRE Trans. on Eng. Man.*, VII (1960), pp. 54–62; R. G. Krohn, 'The Institutional Location of the Scientist and his Scientific Values,' *IRE Trans. on Eng. Man.*, VIII (1961), pp. 133–8; Storer, op. cit., Ch. 5. Barber suggests two more, emotional neutrality and rationality. But it can be argued that both can be subsumed under the norm of organised scepticism. B. Barber, *Science and the Social Order* (1953): emotional neutrality is 'an instrumental condition for the achievement of rationality. . . . it is a necessary component of the moral dedication to the scientific values and methods.' (p. 88).

8. Storer argues that all social systems are organised around the exchange of valued commodities and take steps to ensure that the rewards available are 'system appropriate'. This is essential to maintain the autonomy of the system. Hence, the condemnation of using money to obtain political influence, religious salvation, or sex, op. cit. (1966), Ch. 3.

9. See A. Etzioni, *Complex Organizations* (1961), for an elaboration of this notion of three types of compliance.

10. Thus, in Storer's view, the norms of science are derived from the scientist's interest in the reward, i.e. competent response, whereas, Storer claims, that Merton 'seems to be saying that the reward and the norms are separate consequences of a deep devotion to the advancement of knowledge'.

Storer's apparent uncertainty about what Merton is, in fact, saying is understandable particularly since Storer wished to make a point of distinction between himself and Merton. However, it is a little difficult to believe that Merton would subscribe to the view that scientists uphold the norms of science, because they are devoted to the advancement of science and see the relevance of the norms for this end. Firstly, Merton as Storer points out, warns us against confusing the institutional level of analysis and the motivational level. Certainly one can maintain that certain norms are functional for a particular goal without at the same time stating that individuals behaving in that social situation have internalised the goal as their prime point of orientation. Secondly, Merton in making this distinction does not directly apply himself to answer the question: 'Why do individual scientists uphold the norms of science?' But when Merton does discuss individual scientists his position is not entirely dissimilar to that of Storer. In his study of priorities disputes, Merton says, 'Continued appraisal of work and recognition for work well done constitute one of the mechanisms that unite the world of science.' Further, the 'eureka syndrome', the elation in discovery is the prior counterpart to the intense concern over priority. In these few words, Merton is suggesting that the individual's concern in science is more personal and private than Storer's interpretation of Merton, and that this 'eureka syndrome' is similar to Storer's need for 'competent response'. However, it is to Storer's credit that he has made it explicit that the institutional relationship between values and norms need not be the one which motivates the individual scientist to support those norms.

11. N. Storer, op. cit. (1966), pp. 24–5. One further weakness in Storer's argument is the difficulty in explaining differential commitment to the norms of science. He argues that the need for competent response is the independent variable leading to commitment to the norms. To avoid a circular argument, it would therefore be necessary to offer empirical evidence to show that variations in the need for competent response are causally related to variations in commitment to the norms of science.

12. B. T. Eiduson, *Scientists: Their Psychological Worlds* (1962).

13. H. H. Gerth and C. W. Mills, *From Max Weber* (1958).

14. See Chapter 3.

15. Sometimes referred to as role-identity, i.e. that part of the individual's total identity which is relevant to any specific role. On the concept of role-identity, see G. J. McCall and J. L. Simmons, *Identities and Interactions* (1966). For these interaction sociologists, a role identity is the actor's 'imaginative view of himself as he likes to think of himself being and acting as an occupation of a position'. p. 67.

16. Role analysis is, in fact, more complex than this. We need to distinguish, for example, between the actor's definition of his role, and the definitions held by those with whom he interacts (his role-set). Such distinctions will be made as necessary as the analysis proceeds. Morever, the clarity with which roles are defined varies widely and the ambiguities may be considerable. And there is a variety of ways in which conformity may be minimised or avoided. Indeed, it is useful to think of a transactional relationship between the individual and the organisation, in which the actor seeks to maximise his role bargain. See Chapters 5 & 6.

17. E. Goffman, *Encounters* (1961), We return to this point later in the chapter.

18. A latent identity is one which some influential 'group members define

as being irrelevant, inappropriate to consider, or illegitimate to take into account', A. W. Gouldner, 'Cosmopolitans and Locals', *A.S.Q.*, II (1956–7), pp. 281–306, 444–80. An obvious example is when an attractive female student seeks to intrude her femininity into the student-teacher relation so that interactions are structured by the inappropriate sex-roles.

We do not, in fact, intend to use these terms in quite the same way as Gouldner. He fails to distinguish adequately between the self or identity which the individual brings to a role, and the identity inferred by alter from the role adopted, that is, the product of altercasting. We shall refer to the latter as prescribed-identity. On this, see E. Goffman, *Stigma* (1963), p. 2, and G. J. McCall and J. L. Simmons, op. cit. (1966), pp. 139–41.

19. Publication policies in industry will be examined more fully in Chapter 5.

20. For an attempt to differentiate between types of creative intellectual activity by product and audience, see N. Storer, op. cit. (1966), Ch. 5.

21. We are departing here from the usage which has been adopted by many writers on the sociology of science. Storer, for example, refers to science as a profession. But he is forced to distinguish between 'service' and 'non-service' professions (op. cit. 1966, pp. 16–20). To avoid confusion, we think it preferable to reserve the term profession for that group of occupations which *applies* knowledge and skills in the provision of a service. This normally leads to the adoption of specific occupational strategies centring on the exploitation of expertise including control of entry and qualifications, regulation of conduct, control of market situation through scales of fees, etc. It is notable that in this sense, industrial scientists are only weakly 'professionalised'. We return to this point later.

22. These may be specified in a contract of service or in a job specification, as well as by the formal rules of the organisation. Role specifications clearly vary, in their explicitness and discretion, but a certain minimum of compliance is a condition of continued employment.

23. But not those which stem from the simultaneous occupancy of several roles, e.g. academic scientists *and* professional.

24. D. Marvick. *Career Perspectives in a Bureaucratic Setting*, Michigan Government Studies, No. 27, (Ann Arbor; Bureau of Government, Institute of Public Administration, Univ. of Michigan, 1954).

25. This dichotomy was employed by Donald Pelz in his study of performance of scientists.

26. A. W. Gouldner, op. cit. (1957–8). It has also been used by Marcson, op. cit. p. 18; H. A. Shepard, 'Nine Dilemas in Industrial Research,' *A.S.Q.*, I (1956), pp. 52–8. P. Hollis, 'Human Factors in Research Administration,' in R. Likert and S. P. Hays, (eds.), *Some Applications of Behaviour Research*, Science and Society Series (1957), pp. 124–59.

27. T. Caplow and R. McGee, *The Academic Market Place* (1958), p. 85. See also R. C. Davis, 'Factors Related to Scientific Research Performance,' *Interpersonal Factors in Research, Part 1* (Ann

Arbor; Institute for Social Research, Univ. of Michigan 1954) mimeographed.

28. B. G. Glaser, *Organisational Scientists: Their Professional Careers* (1964): 'Cosmopolitan and local as dual orientations of the same scientist emerged in the analysis of a research organisation that emphasised the institutional goal. Since little or no conflict between goals existed, there was no necessity for taking a priority stand or for splitting into groups. Where there is this congruence of goals, a local orientation helps to maintain the opportunity to pursue research and to have a career at a highly prestigious locale: both conditions are thoroughly consistent with the cosmopolitan orientation.' (p. 28).

29. P. M. Blau and W. R. Scott, *Formal Organisations* (1963), p. 71.

30. R. Avery, 'Enculturation in Industrial Research,' *IRE Trans. on Eng. Man.*, VII (1960), pp. 20–4; see also, W. G. Bennis, *et al.*, 'Reference Groups and Loyalties in the Out-Patient Department', *A.S.Q.*, II (1958), pp. 481–500. For a critical note on dichotomous typologies, see L. C. Goldberg, *et al.*, Local-Cosmopolitan: Unidimensional of Multi-dimensional', *A.J.S.*, LXX (1965), pp. 704–17. Selection mechanisms may also operate to bring about a measure of congruence between individuals and organisations. This point is examined more fully in Chapter 4. See also R. G. Krohn, op. cit. (1961). S. S. West, op. cit. (1960).

31. In an earlier paper (*B.J.S.*, 1966), we referred to this type as the 'instrumental' scientist. However, this term may possibly have some perjorative overtones. Moreover, the term 'organisational' scientist appears to be a more natural description for this type of scientist, since he is willing to move out of science into other roles within the organisation.

32. E. Goffman, 'Role Distance', in *Encounters* (1961).

33. For the interpretation of social identity being that identity available for a role incumbent, see E. Goffman, ibid., esp. pp. 106–7; see also his *Stigma* (1963), pp. 2–3. In his sense, the 'me' is the virtual role self and the 'I' is the self conception; these two become fused in a situation of role embracement.

34. op. cit., p. 164.

35. See on this, Barber and Hirsch, op. cit. (1963), Chapters 15–17.

36. See on this, D. J. de Solla Price, op. cit. (1963).

37. In addition to the previously cited works of R. K. Merton, see also F. Reif, op. cit. (1961); Diana Crane, 'Scientists at Major and Minor Universities: A Study in Productivity and Recognition', *A.S.R.*, XXX (1965), 699–714; S. Cole and J. R. Cole, 'Scientific Output and Recognition: A Study in the Operation of the Reward System in Science,' *A.S.R.*, XXXII (1967), pp. 377–90. Naturally, these studies highlight the relationship between recognition and the quality of publications and not their quantity.

38. Mead's major works were published in 1934–9—he died in 1931. On the sparcity of Simmel's academic publications, see L. A. Coser, 'The Stranger in the Academy,' *A.J.S.*, LVIII (1958), pp. 635–41.

39. op. cit. (1965), pp. 226–36.
40. op. cit. (1963), Ch. 3.
41. M. I. Stein, 'On the Role of the Industrial Research Chemists and Its Relationship to the Problem of Creativity' (unpublished mimeographed paper). Italics ours.
42. J. H. Goldthorpe and D. Lockwood, 'Affluence and the British Class Structure,' *Soc. Rev.*, XII (1963), pp. 133–63.
43. The inquiry was by postal questionnaire, plus interviews of a smaller number. For full details of the field work, see Appendix 3.
44. Appendix 3.
45. The price to be paid for autonomy—a move of 200 miles, or a financial loss—may have been considered too high. This group could well therefore have become largely absorbed into the privatised category, making 27 per cent and 34 per cent for the two parts of the sample, and raising the total proportion of the sample falling in the three cells to 93 per cent and 86 per cent.
46. Evidence for this argument is presented in Chapter 4.
47. See N. Ellis, "The Occupation of Science", *Technology and Society*, 1969, for more detailed evidence, published after we had gone into print.

Chapter 3

The Making of a Scientist

Why do some science students become strongly committed to academic science by the time they leave the university, but not others? Is it the influence of some inspiring teacher, or the prevailing ethos in a particular university? Why are some students not susceptible to such influences? Are there predisposing personality factors, or important influences earlier in life?

OCCUPATIONAL CHOICE AND SOCIALISATION

One factor of major importance has already emerged. As we saw when discussing the Swann Report (in Chapter 1), class of degree appears to be a key variable. It is those who get or expect to get good honours who are most likely to choose an academic career. This immediately raises a problem. It is because they are good at their subject that they identify with academic values and prefer an academic career? Is it future career aspirations which leads them to identify with academic values which influences their career choice?

There is considerable evidence which suggests the influence of career on the development of appropriate occupational identities. As Foote and Strauss have pointed out, individuals identify themselves—answer the question 'Who am I?'—in terms of the names and categories current in the groups in which they participate. By applying these labels to themselves they learn who they are and how they ought to behave, acquire a self and a set of perspectives in terms of which their conduct is shaped.[1]

This perspective presents us with a methodological problem. We cannot, in fact, separate occupational socialisation from occupational choice. The latter feeds back and has an effect on the former. Nevertheless, we shall attempt an analytical distinction between the two, but we must bear in mind throughout that occupational choice is not an event that occurs simply at one point in time. Rather, it is an extended process in which options are gradually narrowed, and in which a preference for a future occupational role may influence the socialisation process. In this chapter then, we will concentrate on exploring the influence of occupational choice, and the effect of education. But we will also probe more deeply to see whether we can discover any underlying predisposing factors in personality or previous experiences, in order to explore more fully the motivations underlying any preferences for specific scientific roles.[2] We will then turn in Chapter 4 to investigate the actual process of career choice.

CAREER INTENTIONS, EXPECTED DEGREE,
AND SCIENTIFIC IDENTITY

If the perspective outlined above is correct, then we can expect career intentions to be a key factor in the development of an occupational identity. Currie,[3] for example, has shown in one inquiry that students had a distinct image of the university professor and that those who contemplated an academic career closely conformed to this image. Moreover, students will be aware of the qualities demanded by their chosen careers and are likely to attach different degrees of importance to those attributes which are most relevant.[4] Thus, the fact that students are exposed to the same influences does not mean that all will perceive or evaluate those influences in the same way.

Now we cannot, in fact, demonstrate conclusively from our research evidence that career choice directly influences occupational socialisation. We *can* show that there is a high association between career choice and identity (see Chapter 4). But the direction of causation is problematic. We can show that the public scientist will seek an academic career. But how far the attraction of such a career results in anticipatory socialisation is more difficult to determine.

We found a strong association between expected class of degree, commitment to science, and a preference for an academic career. We know that a good honours degree is a prerequisite for academic work. And we also know that a high proportion of those with good honours in fact choose a university post in preference to industry. Now expected class of degree may affect *both* the degree of commitment to science *and* choice of career. It is reasonable to infer therefore, that it is the realisation that an academic career is within reach which results in a higher degree of commitment to science among those who expect to get good degrees. Our data is consistent with this interpretation.

Expected class of degree does not, in fact, appear to influence commitment to science (Table 3.1). But it *is* associated with preference for a career in a university (Table 3.2). And the choice of university is strongly associated with commitment to science, 51 per cent of those intending a career in a university compared with 26 per cent being public scientists (Table 3.3).

TABLE 3.1

Expected class of degree is not associated with commitment to science

Expected degree	Per cent public scientists
High	35 (29)
Low	28 (27)
	(p = n.s.)

TABLE 3.2

Expected class of degree is associated with preference for a career in a university

Expected degree	Per cent preferring university
High	34 (28)
Low	12 (11)
	(p = ·001)

TABLE 3.3

Preference for university is strongly associated with commitment to science

Occupational preference	Per cent public scientists
University	51 (20)
Other	26 (36)
	$(p = \cdot001)$

However, although this data lends some support for the theory of anticipatory socialisation, there is little doubt that part of the association between identity and occupational choice reflects a preference on the part of scientists in the making for a career which will be reasonably congruent with their needs. As we shall demonstrate more fully in the next chapter, the preference by public scientists for an academic career reflects an assessment of the rewards and constraints provided by the various alternatives. And in this sense, it is identity which influences occupational choice. We cannot, in fact, on the available evidence, assess with any certainty, the relative influence of identity on occupation compared with that of occupation on identity. The most plausible explanation is to argue that each influences the other.

One last point is, however, worth making. A number of manpower reports have argued that the nature of the PhD degree in its present form influences graduate students away from an industrial towards an academic career. Again, we would argue, that the situation is more complex than this. We found that a number of students intended to go into industry, but also intended to take a higher degree. We also found that students who expected to get a good degree were more likely to take a higher degree before going into industry (Table 3.4), and that public and private scientists (Table 3.5) were slightly more likely to take a higher degree than organisational scientists. But in spite of the fact that many who intend a career in industry decide to pursue a higher degree first as a step towards

TABLE 3.4

*Among students preferring industry those expecting
a good degree are more likely to take a higher degree*

Expected degree	Per cent entering industry after higher degree
High	63 (40)
Low	28 (51)

TABLE 3.5

*The more committed scientists are more likely to take a
higher degree before entering industry*

Identity	Per cent industry after higher degree
Public	52 (12)
Organisational	36 (18)

an industrial career, it is still possible that the PhD experience detracts some from their original career intention. Indeed, the fact that these are students who expect to get good degrees, and those who are more academic in their orientation means that it is precisely these who are most susceptible to the seduction of the academic life. Only further research into changes in career intentions during the PhD course can determine the extent to which this actually occurs.

This evidence is clearly capable of considerable discussion and possible alternative interpretations. It could be argued that the pressures to identify with academic science are so strong that they distort the process of occupational socialisation. We could certainly reject the value judgement implicit in the notion that a university *ought* to socialise individuals to the characteristics currently required by industry. It is also possible that the influence of occupation comes too late. By the time the scientist in the making realises that his ambitions to engage in public science is unrealistic, the process of identity formation has gone too far. Such questions await more intensive research. But some evidence on the influence of the university experience on commitment to science is explored in the next section.

THE INFLUENCE OF EDUCATION: THE UNIVERSITY

Much of the blame for the current imbalance between man-power needs and resources has been laid at the doors of the universities. If this situation is to be changed, it is argued, 'it is of great importance to change a widespread belief that academic research is the only respectable outcome of a scientific education'.[6] In particular, there will need to be reforms at the post-graduate stage, since it is this population which is most reluctant to pursue a career in industry. 'What are required are new forms of postgraduate and postexperience education planned in the light of a careful study of the changing patterns of needs for subsequent careers.'[7]

Influence of staff

In seeking to identify the specific influences at work in the university, we turn first to explore the influence of the teaching staff, since is is these who will play a particularly important part in mediating the values of science. But it does not neces-sarily follow that students will identify with those who teach them, and take them as role-models. The characteristics and values which the student brings with him from his previous socialisation will influence his perception and evaluation of those who teach him. In other words, a student is more likely to become identified with the world of science if there is both a high proportion of dedicated scientists, *and* if they are admired. Our empirical data lends support to this view. Among those who admired public scientists on the university staff, a much higher proportion were themselves public scientists (41 per cent compared with 22 per cent.)[8] (Table 3.6). Moreover, there were considerable differences between universities in the

47

TABLE 3.6

Commitment to science and identification with university staff

Admiration for public scientists on university staff*	Type of Scientist			
	Public	Private	Organisational	$n = 296$
High	41	37	22	122
Low	23	41	36	174
$(p = \cdot001)$				

* (Third-year undergraduates)

proportion of final year students who were public scientists (32 per cent compared with 17 per cent). Such differences could not be explained in terms of differences in the intake, although more students at the university with the higher proportion expected to get good degrees, and more were members of peer groups dominated by public scientists (see below). A possible explanation is that the university producing more public scientists had a higher proportion of such scientists on its staff.[9]

University peer group

Many sociologists have pointed out that peer groups in schools and colleges are important socialising agencies.[10] One influential theme in this area of study is that membership of a student peer group will have a detrimental effect on academic performance, since, it is argued, the core values of the peer group are those of the 'youth' culture which are antagonistic to adult values.[11] This view has, however, been challenged by more recent researches, which suggest that youth culture does not dominate all the perspectives of young people, but is a set of values and beliefs related primarily to age-group activities—to modes of dress, speech, and leisure, rather than a distinctive set of beliefs about the value of work, education and marriage. Following such criticisms therefore, we need to discover the actual content of the values mediated via the peer group. We cannot assume that these are necessarily anti-academic or anti-adult. Moreover, we can expect a student to try to join a peer group whose values reflect his own. The student who is already predisposed to become committed to science is therefore likely to join a peer-group of like-minded students. Such a group would reinforce the tendency to identify with the public scientists on the staff and would play a part in mediating their values.

Again, our data lends some support to such an interpretation, although the association between being a member of a peer group dominated by public scientists and being a public scientist is not a strong one. When, however, we combine having a high admiration for public scientists on the university staff with being a member of a public scientist dominated peer group, we find that 50 per cent of this group are public scientists and only 15 per cent are organisational[12] (Table 3.7).

TABLE 3.7

Commitment to science, membership of peer group, and identification with university staff

	Percentage who are:		
	Public scientists	Organisational scientists	$n = 278$
Both admire public scientists on staff and belong to public scientist dominated peer group	50	16	52
Either, but not both	27	30	99
Neither	25	35	127
	($p = \cdot 02$)		

BIOGRAPHICAL FACTORS

The influence of the university on the degree of commitment to science is considerable. But it can explain only a part of the variance. Why, for example, do some students admire public scientists and join peer group dominated by public scientists while others do not? And why do some become public scientists in spite of the lack of such influences? Such considerations lead us to search for predisposing factors, antecedent to entry to the university.

Birth order, childhood isolation and independence

Since the first known data appeared in Sir Francis Galton's *English Men of Science*, published in 1874, the relationship of birth order to achievement has been repeatedly examined.[13] Although many of these studies have been primarily concerned with birth order and achievement in fields of human creativity there have been some studies which have explored the influence of ordinal position on eminence in science.[14] The evidence is that first-born and only children are indeed over-represented.[15] But there remain two crucial problems to which few sociologists have successfully addressed themselves. Firstly, is the high proportion of first-born or only children a characteristic of eminent scientists rather than other types of scientists? Previous studies have been more concerned with a comparison with the general public. Secondly, if there is an association, how do we explain this?

Numerous explanations have been offered of this observed association. The difficulty is that most of them are post factum, and little rigorous evidence has been produced to test the explanations, by exploring contingent propositions that can be deduced from them.[16] Two alternative theories about the relation between high achievement and ordinal position are particularly worth exploring: (i) the isolation hypothesis, and (ii) the independence hypothesis.

The isolation hypothesis was probably first put forward by Faris.[17] He thought that only children and those who were first-born would tend to experience periods of isolation from others which would not occur if they had other sibs with whom they could interact. In these periods, Faris argued, the child would search out rational courses of action rather than be dependent upon traditional ones handed down by parents or other sibs. This comparatively more rational element in the child's life would also lead to a more consistent and organised childhood. It was these conditions which Faris argued, were necessary for the development of scientific competence.

Recent studies certainly support the view that eminent scientists experience childhood isolation. Eiduson[18] for instance, reported that, 'most scientists experience in their childhoods periods of isolation either stimulated by personal needs, or forced by physical or psychological circumstances, during which they turned or returned to their own resources for solace and amusement, experimented with their abilities, and extended them. Often the experimentation became rewarding and strengthened their interests in using these abilities. However, this reduced interest in "normal" children's games and activities'. Furthermore, Roe[19], in her study of sixty-four eminent scientists, found that they were somewhat 'isolated at times as children. This may have been due to illness, to lack of companionship on an intellectually equal basis, to real or fancied feelings of family superiority. . . . These natural scientists . . . were typically boys who had one or two close friends interested in the same things they were interested in, but who were not inclined to run with a gang, and were quite late in developing dating patterns of various sorts'.

However, although Roe lends support to the first explanation, she attaches more importance to the role of early independence. She considers two hypotheses to explain the over-representation of first-born among high achievers. 'One is that first-born are likely to be over-protected . . . (and thus they) may be compensating in terms of seeking more independence.' The other hypothesis is that eldest sons may be given more and earlier, responsibility than is the case for other children. They have also been spared the discouragement of competition with older siblings. Hence, they are simply continuing an early pattern of independence.[20]

Two inferences may be drawn from these explanations. Firstly, scientists are people who have learned to manipulate things rather than people, and they enjoy this manipulation. Secondly, their relative independence, early isolation and hence their failure to develop social skills, suggests that eminent scientists would prefer situations in which social interactions are minimal. There are a number of studies which lend support to this latter conclusion. McClelland[21] suggests that scientists try to avoid interpersonal contact, and are disturbed by complex human emotions, perhaps particularly by interpersonal aggression. Roe reported that many of her scientists 'had inter-personal relations of a generally low intensity, and showed a much stronger preoccupation with things and ideas than with people'. Schachter's[22] study shows how experiences at university may reinforce early childhood preferences for social isolation. He found that birth order was associated with student popu-larity ratings: that students who were first born or only children were less frequently rated as popular. This lesser popularity, Schachter suggested, probably resulted from having less-developed social skills, itself a function of restricted interpersonal

TABLE 3.8

Public scientists are more likely to have experienced childhood isolation and early independence

	Public scientists Per cent	Private scientists Per cent	Organisational scientists Per cent	$n =$
First/only born	42	33	25*	361
Childhood isolation	44	33	23†	217
Childhood independence	52	25	20†	126

* $p = \cdot01$.
† $p = \cdot02$.

interactions in childhood. Furthermore, involvement in in-tellectual pursuits, he argued, was a consequence of this lack of popularity. A further piece of supporting evidence comes from a study by Davis.[23] In his investigation of graduate students, he dichotomised his sample into 'cosmopolitans' and 'locals', which are similar to our public and organisational types of scientist. He found that the former were able to sustain longer periods of isolation from their student peers when compared to students who were classified as 'locals'. His interpretation of this is that cosmopolitan scientists are less dependent upon gratifications from the immediate social environment. One final item of evidence comes from Sanford.[24] In his study of women students it is reported that high academic achievers revealed that they had suffered from 'early and persistent awkwardness in social relations with peers'.

On the evidence of such studies then, we should expect to find that public scientists are more frequently first-born or only children, that they have experienced comparatively more childhood isolation, and they achieved childhood independence at a comparatively earlier age. Evidence from our study of graduate and undergraduate students in three English universities[25] supports our expectations. We found that a higher proportion of first-born among public scientists than among organisational scientists (42 per cent compared with 25 per cent) and a higher proportion of those who had experienced childhood isolation and early independence (Table 3.8).

Religious beliefs

Over fifty years ago, Leuba[26] showed that scientists and academics were, on average, less religious than were business people, lawyers and bankers. Subsequent studies involving small, unrepresentative samples produced evidence which pointed in the same direction. For instance, Bello[27] examined the religious beliefs of one hundred famous living scientists. His evidence showed a clear falling away of religious convictions, irrespective of original affiliations as indicated by the religious beliefs of parents. Furthermore, religious convictions in this group were less frequent than in the general adult population. Ann Roe also found an association between scientific eminence and lack of religious beliefs in her impressionistic study of sixty-four scientists. Although all but one of these had come from a religious background, 'now only three . . . were seriously active in church'. Most of them were 'not personally concerned over religious matters'.[28] Other writers, such as Ellis, O'Dea, and Kane,[29] have studied the same phenomenon, from a different angle. They have attempted to show that few men with strong religious commitments become major contributors to science.

Recent studies have attempted to explain the observed association. However, despite some close examination, they have failed to produce an explanation which commands a general consensus. Broadly speaking, the various explanations can be placed into one of two categories. These have been summed up by Thalheimer[30] as: '(i) professional training, subsequent academic work experience, and interaction in the academic subcultural setting result in the abandonment of traditional religiosity; and (ii) the academic professions attract a relatively higher proportion of irreligious individuals than do any or most other occupations.' The first type of explanation suggests that there is a basic incompatibility between the ethos of science and religious beliefs, and that intensive exposure to the former erodes belief. Stark[31] found that the more students were exposed to the 'scientific ethos', the less they were religious. Stark has summed up his work in this field by suggesting that 'graduate training, probably at a secular school, is a usual part of the process by which men come to be scientific scholars and it appears that during that process, religion is falling away.'[32] Greeley[33] challenged this conclusion, claiming that his own evidence did not support such an interpretation. Indeed, he suggested that 'many young people who have rejected the traditional religions are attracted to science as a functional alternative for religion'.[34] He would argue then, that it is not exposure to science which causes religious beliefs to decline, since these beliefs have already declined before such exposure. In other words, identification with science meets a need which arises from not being religious.

Thalheimer's most recent studies support the second of these explanations. From his study of over seven hundred academics in one American West Coast state university, he concluded that 'a high proportion of the academicians under investigation had already abandoned traditional religious patterns by the time they began graduate work and that changes in either direction are comparatively infrequent after college. Assuming that this is indeed what takes place, it would seem to follow that professional training and academic teaching and research cannot be considered to play an important role in the process of religious change, except for a relatively small minority of academicians'.[35]

We would subscribe to the view that involvement in science is a functional alternative to religious involvement. Moreover, we would expect to find that not only are scientists as a group less religious than non-scientists, but that public scientists are less religious than those who are less strongly committed. There is already some evidence to support this. For instance, Lazarsfeld and Thielens[36] found that academics in non-religious universities were more productive than academics in religious (i.e. Protestant or Catholic) universities. A similar conclusion could be drawn from the work of Roe and Bello cited earlier. Both studied only famous scientists, i.e. those whose recognition had followed a period of fairly intense publication. More direct support comes from a recent paper published by Vaughan, Smith and Sjoberg.[37] They showed that scientists who have studied a pure science, in this case chemistry, are less religious compared with those scientists who had studied chemical engineering. Furthermore, the nonreligious scientists preferred to work in basic research and/or in an academic location.

Finally, Stark compared the intensity of religious beliefs of 'cosmopolitan' and 'local' scientists, and found that 60 per cent of the latter had strong religious involvements compared with only 35 per cent of the former.[38] Our evidence also supports the view that the rejection of religious beliefs and the different degrees of religious beliefs, compared with those held in an earlier period of one's life, are strongly associated with the commitment to science (Table 3.9).

TABLE 3.9

The less religious are more committed to science

	Per cent public scientists	$n =$
Not religious	45	235
Beliefs decreased	38	84
Beliefs unchanged	31	170
Beliefs increased	27	78

$p = .001$.

Political and social values

Not only are scientists less religious, they are also more radical in their political and social attitudes. Lazarsfeld and Thielens[39]

55

found that the more productive scientists were more likely to support the Democratic Party and to take a generally liberal view on the current civil rights issues of the 1950s, particularly the rights to free speech and belief. Kornhauser[40] also found that scientists were 'more likely to favour the Democratic Party, less likely to favour *laissez-faire*, and less likely to express prejudiced attitudes towards minorities'. Furthermore, 'these attitude differences between scientists and engineers cannot be attributed to differences in the two group's social origins, religion, or parents politics, since they were remarkably similar in respect of these factors'.

Our own evidence also supports the view that commitment to science is associated with political radicalism. Of those in our sample who were in complete agreement with Labour's principles, 63 per cent were public scientists and 12 per cent organisational. By contrast, none who were in complete agreement with Conservative principles were public scientists, but 45 per cent were organisational (Table 3.10).

TABLE 3.10

Public scientists are more radical politically

Political party	Belief in principles	*Per cent* Public	Organisational	(n = 536)
Labour	Completely	63	12	
	Strongly	49	18	
	Not Strongly	28	37	
Liberal	All	41	22	
Conservative	Not strongly	37	25	
	Strongly	33	32	
	Completely	0	45	

One possible explanation of this greater radicalism of scientists is to be found in their early socialisation. There is some evidence to suggest that children from families in which independence is granted early develop a high need for achievement.[41] It is also plausible to argue that such children develop a high tolerance for ambiguity and uncertainty. Relative isolation and independence will have given him early experience in finding his own way in situations, rather than responding to them in terms of 'received' traditional reactions mediated through older siblings. It is then, their early independence rather than the fact that they are scientists which explains their greater radicalism.

Of course, political attitudes are a crude measure of radicalism. Although supporters of the Conservative party are more likely to be opposed to radical changes particularly in certain aspects of the economic organisation of society, it does not follow that Labour voters are necessarily more radical in the sense of being anti-tradition, or more attracted to rational solutions. Nevertheless, the association is sufficiently strong to deserve further exploration.

Both the rejection of religious belief and political radicalism may be expressions of a more general underlying factor— 'anti-traditionalism'. Stark's survey of comparative evidence led him to suggest that to the degree that men seek to alter the existing stratification arrangements, 'they are likely to have turned away from the prevailing religious institutions of their societies'.[42] It is such anti-traditionalism which is expressed in science in the form of organised scepticism.

THEORETICAL PERSPECTIVES ON THE SOCIALISATION PROCESS

The above discussion has demonstrated the existence of a number of factors which play a part in the making of a scientist, and in particular, those which are associated with a high degree of commitment to science. The explanation so far, however, is little more than a listing of variables, and does not amount to a coherent theory.

We can improve our explanation somewhat by demonstrating that the effects of the discrete variables are cumulative. Although each factor can account for only a small part of the variance, taken together, we can account for a considerable amount of the total variance. If, for example, we control for social class origins, and combine biographical factors, university reference group and occupational preference we find that 100 per cent of working class students with all three predisposing factors are public scientists (Table 3.11). Similarly, for those where all

TABLE 3.11

Relation between biographical factors, university influences, career choice, and commitment to science among working-class undergraduates

Biographical factors*	University reference group	Percentage in cell who are public scientists†		
		(a)	(b)	(c)
High	Public	100	50	71
	Other	44	18	29
Low	Public	56	33	41
	Other	31	3	18
				$n = 139$

* 'High' scored as two or more of being first/only born, experiencing childhood isolation, being non-religious.

† (a) Prefer academic career
(b) Others
(c) Total.

the aspects of socialisation are negatively associated, 3 per cent are public scientists. Thus, it is possible to account for a very large amount of the variance in terms of the factors which we have isolated.

But such an approach is theoretically crude. A more satisfactory theoretical perspective would be to achieve a higher order integration of a number of variables, so that the effect could be explained with parsimony and elegance by reduction to a few simple theoretical propositions. We shall argue, that two major processes can be identified in the socialisation of scientists. The first of these theoretical solutions argues that becoming a scientist is a means of solving problems resulting from the experience of social marginality.

Social marginality and commitment to science

In the early formulation of the theory of the marginal man[43] no distinction was made between being in a situation of marginality, and the subjective experience of marginality. The situation of being caught between two cultures was seen as necessarily generating personality conflict.[44] It was not, however, until the 1940s that it was pointed out that the relation between situation and personality is empirically problematic rather than axiomatic.[45] More recently, most studies of marginality have consistently drawn a distinction between the marginal *situation* and the individual's *response* to it, seeing the subjective definition of the situation or *experience* of marginality as the link between the two.[46] This, while representing a major advance on the earlier formulations, still leaves to be discovered those intervening variables which transform situation into experience, and experience into predictable response.[47] For the moment, we will concentrate on the problem of explaining the differential socialisation of working-class science students.

We define the situation of marginality for the present purpose as that in which an actor who has biographical roots in one social world, currently operates in another. These two social worlds must be sufficiently dissimilar for there to be at least a possibility of conflict between their norms.[48] The concrete case is that of the working class student who finds himself in a social milieu, the university, which is dominantly middle class.[49] The working class university undergraduate is thus in a situation which is *by definition* marginal.

Yet whether this situation is experienced as one of marginality, whether there is a sensation of discontinuity between biographical and present self, depends in large part on the actor's attachment to his biographical world. A student who has strong affective ties with his family and class or origin may well subscribe to the social and political values of that class. But the experience of higher education will have exposed him to certain ways of acting, thinking, and even feeling, which generate an awareness of difference and separation from his class of origin. Thus the individual is estranged from his former family and peers. He is exposed to 'a process involving double risk: alienation from erstwhile friends and fami y and rejection by ingroup members at the higher level'.[50]

Under this set of circumstances—cut off from one social class and either unable or unwilling to become integrated with another—the individual may well *experience* an intense form of marginality, a severe crisis of identity.[51]

The outcome of an individual student's attempt to solve such a crisis will depend on the interplay of a number of factors. Firstly, it will depend on the intensity and scope of the crisis. This in turn will be affected by the three biographical variables which we have already seen to be associated with commitment to science, namely birth order, social isolation in childhood, and declining religious conviction. There is a fair degree of consensus that first-born and only children may have difficulty in engaging in situations which require them to employ social skills, and that they may prefer to deal with things rather than people. We should, therefore, expect such a person to attempt to solve the crisis of identity stemming from the experience of marginality in a way which minimised the importance of social skills.

Strong religious beliefs may also be relevant here. Religious beliefs can help an individual to cope with the crisis of identity resulting from the experience of marginality which may result from upward mobility. Religion may act as a 'bridging culture',[52] providing a link between past and present social worlds.[53] That is to say, the individual may solve the question of 'Who am I?' by reaffirming his acceptance of a religious ideology, and this religious identity may take priority over the class-related and conflicting options between which non-religious marginal men may still hover.[54]

However, religious involvement is not the only, or indeed, the most frequent, solution to the experience of marginality, for two reasons. In the first place, it has little obvious relevance to the immediate academic situation. In the second place, religious convictions might themselves generate new tensions between the claims of science and religion. The availability of other solutions to the identity crises depends largely on the individual's experience at school—whether science was a good subject for him, giving him feelings of success, and whether he identified with the science teachers—and his reference group at university.

Thus far, it has been suggested that the marginality experienced by some upwardly mobile students may lead them to search for a new identity. Goffman[55] has pointed out that in any role a virtual self awaits the role player. If he embraces the role, the individual may become embraced by it, and it can thus provide for him a complete ego-identity. Hence the solution to the crisis of identity, which is contingent on the experience of marginality, will depend partially on the roles and identities which are available as options.[56] In fact, the working class students in our sample embraced the role of dedicated scientist far more frequently than did middle class students. There are several possible reasons for this. Firstly, the image of the scientist tends to be a classless one, and hence it can be embraced by a working-class student without involving painful denials of his past. Secondly, the role of scientist is one in which skill in dealing with things rather than people is premiumed. Thirdly, the relevant role models are highly visible to the students.[57] The educational experience of these students will be one in which the achievements of 'great men' have often been cited. As Eiduson has said, 'scientists are idols-oriented';[58] the rarity and hence quasi-sacred nature of 'great thought' is stressed. Fourthly, the bright paint of these idols enhances not only their visibility, but also their desirability, for the main reward of science is recognition. It is through emulation of these role-models that the internalisation of the culture of science is effected. As the student-scientist learns to play the role of scientist, he acquires the motivations of a scientist— motivations directed towards recognition—and ultimately may embrace the scientist's identity.[59] Finally, the dominant values and norms of science are, like those of religion, sufficiently oecumenical[60] to act as anchorage points for other life areas.

To sum up, we may hypothesise that a working-class student who experiences marginality is more likely to become committed to science if (a) he is first-born and has experienced childhood isolation; (b) is non-religious; (c) has identified with public scientists at university. Our empirical evidence is consistent with this (Table 3.11, col c). Indeed, of those students who prefer an academic career, 100 per cent are public scientists.

Social isolation and commitment to science

The argument thus far has been confined to students from working-class backgrounds experiencing marginality and a crisis of identity. But for the middle-class student, such a process is unlikely. For him, it is his failure to develop social skills and his early investment in ideas and things rather than people which is most likely to have channelled him into science. Such early experiences will make social adjustment at a university more difficult and generate a problem of identity which is in many ways similar to that of the marginal working-class student. But for the middle-class student lacking social skills the problem is that he cannot easily obtain reinforcement for his identity from interactions with others. Yet it is from the responses of others that we build up our self images: in Cooley's term—our 'looking-glass' self. The individual who finds it difficult to relate in personal terms, will find it easier to obtain confirmation for his identity in less face-to-face forms of recognition. For such men, public science, academic or fundamental research, offers rewarding roles relatively undemanding in terms of dramaturgical skills.

For our middle-class students we found that a higher proportion of those who were public scientists had experienced childhood isolation (41 per cent compared with 27 per cent), and more lacked social skills (46 per cent compared with 17 per cent). Both the marginal working-class student, and the middle-class student are characterised by a preference for roles which do not demand personal skills, both have similar university reference groups and occupational preferences. They differ only in the steps by which they come to such choices. For the marginal student, it is the result of his being upwardly mobile and experiencing this as a problem of identity. The middle-class student too, faces a problem of identity, but for him it springs less from marginality than from his search for a reinforcement and recognition of the self which does not depend on face-to-face relationships and interpersonal skills. And for both, science offers a role which can provide not only recognition and identity, but a way of life.

If this theory is correct, then it can be argued that for the working-class student, the crisis of identity which springs from his lack of dramaturgical skills is compounded by the experience of marginality. If this is so, then it can be hypothesised that, if we control for other socialising factors, working-class students are more likely to become committed to public science.

SUMMARY AND DISCUSSION

We are now in a position to attempt a tentative answer to some of the questions raised at the beginning of this chapter. There is little doubt that the influence of the university is considerable. Students are strongly exposed to academic values. The most esteemed model of the scientist available for emulation is the academic scientist. Small wonder that some are proselytised. But the situation *is* complex. Realistic estimates of future career chances have some considerable influence. The better students are more likely both to prefer university to industry and consequently to become academic scientists. But what if industry were to become more attractive to the good honours graduate? The evidence on anticipatory socialisation suggests that in such circumstances, fewer good graduates would acquire the identity of academic scientist. There is a strong case for arguing that future career aspirations play a part in the process of identifying with academic values, though how strong a part we cannot say. And the evidence on the characteristics of those who take higher degrees before entering industry also lends support to the argument of the Swann Report that many such students may be particularly susceptible to the attractions of a university career. But firmer conclusions on such points must await a more detailed analysis of the motives behind career choice to be explored in the next chapter.

The influence of the university alone cannot explain why only some become committed to academic values. Undergraduates bring with them to the university predispositions which push some more strongly than others to identify with the values and norms of academic science. For some, being a scientist differs in significant ways from being a clerk or a salesman. The role of academic scientist provides him with an identity which he can *embrace*; he *is* a scientist. It is in this sense that science is a *vocation*. And the recognition which he receives from the 'invisible college' is essential for the constant reinforcement of his identity. In answer to the question, 'Who an I?', he can answer, 'I am a scientist whose worth is recognised by the community of science which transcends the narrow and petty boundaries of class, or nation, or race. This is the world to which I belong and owe my allegiance, to the world of Pasteur, Rutherford, and Niels Bohr.'

But such men are not typical of those who are awarded bachelors' degrees in science at universities. Nor are many likely to find their way into industry. It is to this question—the process of occupational choice and selection that we now turn.

NOTES

1. 'The Elements of Identification with an Occupation,' H. S. Becker and J. Carper, *A.S.R.*, XXI (1956), pp. 341–8.
2. The need for such research has been emphasised in a number of recent reports. The report of the Committee on Manpower Resources for Science & Technology, *A Review of the Scope & Problems of Scientific & Technological Manpower Policy* (Cmnd. 2800, 1965), included among the areas requiring further investigation 'the identification of motivations character and abilities relevant to science and technology, and the factors which govern the choice by individuals of a scientific or technological career'. (ibid, p. 5). See also *Interim Report of the Working Group on Manpower Parameters for Scientific Growth* (Cmnd. 3102, 1966).
3. I. D. Currie, *et al.*, 'Images of the Professor and Interest in the Academic Profession,' *Sociology of Education*, XXXIX (1966), pp. 301–23.
4. K. W. Back, *et al.*, 'Public Health as a Career in Medicine: Secondary Choice with a Profession,' *A.S.R.*, XXIII (1958), pp. 533–41. They show that public health is a deviant medical career and that students choosing this differentially evaluate job characteristics, receive different treatment and encouragement from university staff, and receive low sociometric ratings from university peers. See especially, M. J. Huntington, 'The Development of a Professional Self-Image' (pp. 179–88); R. C. Fox, 'Training for Uncertainty' (pp. 207–44) in R. K. Merton (ed.), *The Student Physician* (1957); H. S. Becker, *et al.*, *Boys in White* (1961); E. L. Quartantelli, 'Faculty and Student Perception in a Professional School,' *Sociological and Social Research*, XLIX (1964), pp. 32–46; H. S. Becker and B. Geer, 'The Fate of Idealism in Medical School,' *A.S.R.*, XXIII (1958), pp. 50–6. See also H. S. Becker and J. Carper, op. cit., *A.J.S.*, LXI (1956), pp. 289–98; A. K. Fisher, 'The Indoctrination of Dental Students with a Professional Attitude,' *J. of Dental Education*, XXIV (1960), pp. 38–41.
5. The classification of private and organisational scientists used in Chapters 3 and 4 differs slightly from that used in the remainder of the study. In these chapters, the thirty-six students scoring L L H in Table 2.2 have been categorised as 'private', following the classification used by Dr Box in his thesis. In subsequent chapters, L L H are

included as organisational scientists, following the typology established in Chapter 1, except where the distinction between L L H and L L L is explicitly stated.

6. *Interim Report* (1966), p. 25.
7. ibid. Such views have, of course, been reiterated in subsequent reports, notably the Swann Report.
8. Data presented in Steven Box, *Chemistry Students and Professional Values* (unpublished PhD thesis, London, 1966).
9. Caplow and McGee have argued market forces result in the more highly productive (i.e. public) scientists concentrating in a few high status universities, which are then in a stronger position to attract more public scientists, who are likely to publish early in their careers and thus become more mobile. Although these above remarks are based on American researches, there is little reason to doubt that a similar process is going on in this country. Available evidence suggests the existence of both a hierarchy in organisations of higher education, and a gravitation of public scientists into well known institutions. See, Caplow and McGee, *The Academic Market Place*; P. F. Lazarsfeld and W. Thielens Jnr., *The Academic Mind* (1958); Crane, op. cit. See also L. Hargens and W. O. Hagstrom, 'Sponsored and Contest Mobility of American Academic Scientists,' *Soc. of Ed.* (1967), pp. 24–38.
10. For examples, see J. C. Coleman, *The Adolescent Society* (1961); E. C. Hughes, *et al.*, 'Student Culture and Academic Effort' in N. Sandford (ed.), *The American College* (1962); W. L. Wallace, 'Peer Influences and Undergraduates: Aspirations for Graduate Study,' *Sociol. of Ed.*, XXXVIII (1965), pp. 377–92; C. N. Alexander and E. Cambell, 'Peer Influences on Adolescent Aspirations and Attainments' *A.S.R.*, XXIX (1964), pp. 568–75; E. L. McDill and J. C. Coleman, 'Family and Peer Influences in College Plans of High School Students,' *Sociol. of Ed.*, XXXVIII (1965), pp. 112–26.
11. The idea that youth culture is antagonistic to academic achievement can be traced back to the early pronouncements by Talcott Parsons on this subject. See his 'Age and Sex in the Social Structure of the United States,' in *Essays in Sociological Theory* (1954, rev. ed.).
12. The influence of secondary education was not explored in any detail, but we did find that pupils who admired their science teachers or who felt particularly successful at science were more likely to become public scientists (significant at the 1 per cent level). On the crucial influence of mathematics in the 'choice of science', see the Dainton report. On the influence of teachers, see also R. A. Ellis and W. C. Lane, 'Structural Supports for Upward Mobility,' *A.S.R.*, XXVIII (1963), pp. 743–56. The authors suggest that where real parents because of their lack of education and knowledge become ineffectual guides, teachers act as parent surrogates.
13. For a summary of these studies, see W. D. Altus, 'Birth Order and Its Sequelae,' *Science*, CLI (1966), pp. 44–9.
14. For example, see J. K. Cattell and D. K. Brimhall, *American Men of*

Science (3rd Ed., 1921); S. S. Visher, 'Starred Scientists 1903–43,' in *American Men of Science* (1946).

15. See the recent study by S. Steward West, 'Sibling Configurations of Scientists,' *A.J.S.*, LXVI (1961), 268–74.
16. Bayer has recently attempted to reduce the ambiguities surrounding this problem. His survey of the explanations which have been offered led him to conclude that they can be reduced to three generic types: (i) intra-uterine and prenatal influences; and (ii) economic interpretations of achievement differences, and (iii) ordinal position differences in socialisation. His survey led him to conclude that we can safely reject the second type of explanation that relies on the proposition that parents can afford to make economic sacrifices for the first-born child, but not for subsequent children. A. E. Bayer, 'Birth Order and Attainment of the Doctorate: A Test of Economic Hypotheses,' *A.J.S.*, LXXI (1966), pp. 540–50.
17. R. E. L. Faris, 'Sociological Causes of Genius,' *A.S.R.*, V (1940), pp. 689–99.
18. B. T. Eiduson, *Scientists: Their Psychological World* (1962).
19. A. Roe, *The Making of a Scientist* (1952).
20. A. Roe, 'A Psychological Study of Eminent Psychologists and Anthropologists, and a Comparison with Biological and Physical Scientists,' *Psych. Mono.*, LXII (1953), p. 47
21. D. C. McClelland, 'On the Psychodynamics of Creative Physical Scientists' in H. E. Gruber, *et al.*, *Contemporary Approaches to Creative Thinking* (1962), p. 144.
22. S. Schachter, 'Birth Order, Eminence and Higher Education,' *A.S.R.*, XXVIII (1963), pp. 757–68, and 'Birth Order and Sociometric Choice,' *J. of Abn. and Sociol. Psy.*, LXVIII (1963–4), pp. 453–6.
23. J. A. Davis, 'Locals and Cosmopolitans in American Graduate Schools,' *Int. J. of Comp. Sociol.*, II (1961), pp. 212–23.
24. N. Sanford, 'Motivations of High Achievers' in D. D. David (ed.), *The Education of Women* (1959).
25. 567 students completed a mailed questionnaire in 1965. For a more detailed analysis of this data, see S. Box, op. cit., 1966, and S. Box and J. Ford, 'Commitment to Science: A Solution to Student Marginality,' *Sociology* (Sept., 1967).
26. J. H. Leuba, *The Belief in God and Immortality* (1921), pp. 219–87.
27. F. Bello, 'The Young Scientists,' in P. C. Obler and H. A. Estrin, *The New Scientist* (1962), pp. 62–81.
28. A. Roe, op. cit. (1953).
29. J. T. Ellis, 'The American Catholic and Intellectual Life,' *Thought*. XXX (Aug., 1955). T. F. O'Dea, *American Catholic Dilemma* (1958). J. J. Kane, *Catholic-Protestant Conflicts in America* (1955).
30. F. Thalheimer, 'Continuity and Change in Religiosity: A Study of Academicians,' *Pac. Soc. Rev.*, VIII (963), pp. 1–18.
31. R. Stark, 'On the Incompatibility of Religion and Science: A Survey of American Graduate Students,' *J. for the Scientific Study of Religion*, III (1963), pp. 1–18.

32. ibid., pp. 14–15.
33. A. M. Greeley, 'Comments on Stark's "On the Incompatibility of Religion and Science",' *J. for the Scientific Study of Religion*, III (1964), p. 239.
34. A. M. Greeley, 'Influence of the Religious Factor on Career Plans and Occupational Values of College Graduates,' *A.J.S.*, LXIX (1964), pp. 658–71.
35. op. cit. (1965), p. 107. However, the sample was small and the response low; the author warns, these 'findings should be viewed as suggestive and not necessarily representative of the entire faculty population from which responses were obtained.' p. 102.
36. op. cit. (1958).
37. T. R. Vaughan, D. H. Smith, and G. Sjoberg, 'The Religious Orientations of American Natural Scientists,' *Social Forces*, XLI (1966), 519–26.
38. R. Stark and C. Y. Glock. *Religion and Society in Tension* (1965), Ch. 11.
39. op. cit.
40. W. Kornhauser, op. cit. (1962).
41. D. C. McClelland, *The Achieving Society* (1961).
42. R. Stark and C. Y. Glock, op. cit. (1965).
43. R. E. Park, 'Human Migration and the Marginal Man,' *A.J.S.*, XXXIII (1928), pp. 881–93. E. Stonequist, *The Marginal Man* (1937).
44. Thus, for example, T. Sellin, *Culture Conflict and Crime* (1938), pp. 63–70, and S. Kobrin, 'The Conflict of Values in Delinquency Area,' *A.S.R.*, XVI (1951), pp. 653–61, where he developed the idea of isomorphic mental conflict.
45. A. W. Green, 'A Re-examination of the Marginal Man Concept,' *Social Forces*, XXVI (1947), pp. 167–71.
46. A. Antonovsky, 'Towards a Refinement of the Marginal Man Concept,' *Social Forces*, XXXV (1956), pp. 57–62; H. F. Dickie-Clark, '*The Marginal Situation* (1966). J. W. Mann, 'Group Relations and the Marginal Personality,' *Human Relations*, XI (1948), pp. 77–92; A. C. Kerckhoff and T. C. McCormick, 'Marginal Status and Marginal Personality,' *Social Forces*, XXXIV (1955), pp. 48–55; R. H. Turner, *The Social Context of Ambition* (1964), pp. 1–18.
47. W. J. Talcott Parsons and E. A. Shiles (eds.), *Towards a General Theory of Action* (1962), pp. 279–361.
48. See Turner, op. cit., pp. 8–9.
49. This is not to say that students necessarily think of themselves as middle class, but rather that the university can be seen as a transitional zone prior to middle class status. For a discussion of this, see J. Abbott, 'The Concept of Motility,' *Soc. Rev.*, XIV (1966), pp. 153–61.
50. Turner, op. cit., pp. 12–15.
51. Throughout this paper the concept of identity is employed in accordance with the usage of the Interactionist and New Chicago schools. See, for example, E. Goffman, *Stigma* (1963), especially p. 2, and 'Role Distance' in *Encounters* (1961). P. L. Berger, *Invitation to*

Sociology (1963), especially Chap. 5; A. L. Strauss, *Mirrors and Masks* (1959); A. M. Rose, *Human Behaviour and Social Processes* (1962); T. Shibutani, *Society and Personality* (1961).

52. 'By bridging culture, we mean that set of values which are salient to the individual and enable him to experience personal continuity between two social status which are not continuous, or which have many point in conflict.'

53. This might well have been stated in the language of cognitive dissonance theory. One mode of reduction of dissonance is that of denying the salience of one of the elements in the relationship. Cf. L. Festinger, *A Theory of Cognitive Dissonance* (1957).

54. Indeed, it may not be too fanciful to compare the community of saints from which the relatively isolated religious mystic seeks an identity, with the scientific community, which frequently claims an almost equally monastic dedication.

55. E. Goffman, op. cit. (1961), Chap. 3.

56. Clearly the role of student is itself one which could be embraced by an individual. Intense involvement in what Burton Clark has called the collegiate subculture in the university may reflect such embracement. However, although the student identity may present a temporary solution to the marginality crises, it cannot be an abiding one. There are, of course, other possibilities; for example, for women students there are those identities to be found embracing the roles of wife and mother.

57. B. G. Glaser, *Organizational Scientists* (1964), p. 129. See, for a discussion of the high visibility of scientist role-models, Fred Reif and A. L. Strauss, 'The Impact of Rapid Discovery on the Scientist's Career,' *Social Problems*, XII (1965), pp. 297–311.

58. B. Eiduson, op. cit., pp. 167–8. See also, L. Kubie, 'Some unsolved problems of the Scientific Career' in Barber and Hirsch, op. cit., pp. 199–201.

59. See Glaser, op. cit., p. 130, and also O. Klapp, *Heroes, Villains and Fools* (1962), pp. 18–24.

60. See T. Parsons, 'Some Aspects of the Relation between Social Science and Ethics' in Barber and Hirsch, op. cit., pp. 590–5, and V. K. Dibble, 'Occupations and Ideologies,' *A.J.S.*, LXVIII (1962), pp. 229–41.

Chapter 4

Career Choice

Choice of career is not an event which can be located at one point in time. It is a process which stretches back into childhood, where basic personality characteristics begin to be formed. These in turn will channel the broad direction of choice. And, with the passage of time, further influences and decisions will gradually narrow the range: childhood isolation giving a push in the direction of preferring things to people; while university experience can reinforce an interest in academic research. Moreover, as the range of possible choices narrows, future career prospects begin to cast their shadows before them—and there is an element of anticipatory socialisation—of becoming more like what the job requires. So, as we have seen, the processes of occupational socialisation and selection interpenetrate. But for analytical purposes, we concentrate in this chapter on the actual choice of first job, against the background of the given characteristics which the individual scientist possesses at the point of career choice.

A SOCIOLOGICAL THEORY OF
OCCUPATIONAL CHOICE[1]

The task we have set ourselves, of trying to explain why scientists choose one job rather than another assumes that the process of choice is purposive and not merely adventitious. This is an assumption which we hope to demonstrate is, in fact, justifiable—that clear choices are made according to identifiable criteria. But our task is simplified if we maintain the analytical distinction between choice and socialisation—a distinction which has by no means always been clear in the literature.[2] There may well be fortuitous elements in the total process by which individuals are channelled towards one occupation rather than another—accidents of birth, of experience, and even genetic factors over which they can have no control. Only those born with the genetic potential of above average height can choose a career in the police. But in this chapter, we are focusing on the options available and the choices made by those whose socialisation is substantially complete: who *are* particular kinds of persons, with particular characteristics, and therefore preferences for certain kinds of satisfactions. We are arguing that in making a choice, the individual will seek a career which he sees as desirable—as one which will be rewarding—as one in which he will have the best chance of realising the various needs, hopes, and expectations which at the moment of choice he believes to be important. Of course, such choices may be made on the basis of imperfect knowledge, or even distorted and inaccurate information, both about himself, and about the opportunities and rewards which alternative careers offer. But it will still be purposive and rational in the sense that he will act on the best available information. And in terms of the concepts which we have previously used, this means that he will seek to optimise the congruence between his identities, and the various role options which are *available* to him.[3]

Our problem can be re-stated then in the form of two testable propositions:

1. 'In choosing between alternative occupations, a person will rank the occupations in terms of the relation between his values and the perceived characteristics of the occupation; the higher the coincidence between the characteristics and his values the higher the rank.'

2. 'The higher a person perceives the probability that he will obtain employment in the higher-ranked occupation, the more likely he is to choose that occupation.'[4]

THE PERCEPTIONS OF OCCUPATIONS

We have already established that by the time they reach their final year of undergraduate study, science students differ considerably in their values—specifically in the extent to which they have identified with the values of academic science (*public* compared with *organisational* scientists). How then, do they perceive the rewards which are available in different types of occupation? If we concentrate on the areas of most significance for this study, we find marked differences between their perceptions of work in universities and industry (Table 4.1).[5]

TABLE 4.1

Percentage of students* perceiving various aspects of employment as better than adequate

	Industry Per cent	University Per cent
Salaries	75	24
Holidays	38	63
Freedom to publish	19	94
Freedom to choose projects	10	78
Technical equipment	82	57
Freedom to choose work colleagues	8	36
Social and welfare facilities	54	60
Career ladder for scientists	26	53
Location and surroundings	18	47

* Third-year chemistry undergraduates plus postgraduates.

Most thought that both salaries and technical equipment were better in industry.[6] But the universities were thought to allow more freedom to publish, to choose research projects and colleagues, and more opportunities for an adequate career in science. There was general agreement too, that the universities were better situated geographically and gave more adequate holidays.

It is not our immediate concern here whether the different perceived characteristics of industry and the universities are true reflections of the actual situation. It is how such occupations are perceived by undergraduates which determines their preferences. But it is worth noting that the characteristics of work in these different locations described by scientists in Kelsall's survey and reported in the Swann Report do correspond closely with the differences described here. Kelsall found,

71

for example, that scientists in industry rated their jobs relatively low on opportunities for intellectual development, and scope for initiative and freedom to develop ideas.[7]

Perhaps the image of conditions varies according to future career intentions? Perhaps those who hope to work in the universities see them in a more favourable light? To test this, we controlled for career intentions. But there were not significant differences in the images which students had of industry

TABLE 4.2

Occupational preference does not influence
perception of employment conditions

Job condition	Percentage perceiving job conditions as more than adequate*					
	Industry			University		
	Preferred employment					
	Ind. $n = 129$	Univ. $n = 67$	Per cent diff.	Ind.	Univ.	Per cent diff.
Salaries	81	72	+9	23	25	− 2
Holidays	32	31	+1	77	56	+21†
Freedom to publish	22	13	+9	95	98	− 3
Freedom to choose projects	6	6	—	76	79	− 3
Technical equipment	80	83	−3	56	60	− 4
Social/welfare facilities	53	52	+1	60	60	—
Freedom to choose colleagues	10	6	+4	33	30	+ 3
Career ladder for scientists	24	17	+7	62	58	+ 4
Geographical location, etc.	20	27	−7	56	49	+ 7

* Includes both graduate and third-year undergraduate students.
† $x^2 = 9.65$; $df = 1$; $p = .01$.

or university regardless of future plans—with two small exceptions. Those who intended working in industry had a more favourable view of the adequacy of the holidays it allowed (Table 4.2), and public scientists who intended to work in industry had a different perception from those who did not.

THE INFLUENCE OF IDENTITY ON CHOICE

We cannot then explain occupational choice in terms of any differential perception of conditions. It is to the *significance* of such differences for the individual that we must turn for an explanation. Such differential evaluations will be related to the students' ideal for his future role.[8] We have defined public scientists as those who attach relatively more importance to freedom to publish, and less to salaries. Moreover, we would expect them to choose to work where the conditions they believed to be important could be found. Thus, we would expect public scientists to prefer higher education and organisational scientists to prefer industry. Our findings support such expectations. In our sample of undergraduate chemists[9] 36 per cent of the public scientists preferred a job in higher education compared with 13 per cent of the organisational scientists (Table 4.3). A similar picture emerges for graduate students. To this quantitative evidence can be added some qualitative depth from the students' comments. For example, one student classified as a public scientist gave the following as his reasons for preferring to work in higher education after he had completed his PhD: 'Universities are, in my opinion, the only places where there is: (i) freedom to publish; (ii) a reasonably good amount of equipment; (iii) freedom to decide on research projects; (iv) no "clocking-in" or "out", i.e. freedom to decide when to work, how much one wishes to do on a particular day; (v) less "red-tape"; (vi) an international circle of acquaintances. All at the same place.'

Another student, this time one classified as a private scientist, gave the following for a similar employment preference:

'The life of a university lecturer appeals to me because of the

73

opportunity of doing a job I like, in my own time with *no pushing from above*. I also consider the holidays as being good, again being free within reasonable limits, to choose when to

TABLE 4.3

Scientific identity is strongly associated with occupational preference

Preferred Employment Location	Type of Scientist			
	Public	Private	Organ-isational	$n =$
Undergraduates				
(1) Industry now	11 (18)	27 (33)	14 (36)	52
Industry later*	12 (21)	19 (32)	8 (21)	39
Government	4 (7)	5 (6)	1 (3)	10
Teaching	9 (16)	18 (21)	10 (26)	37
Higher education	20 (36)	14 (17)	5 (13)	39
n (100%) =	56	82	39	177†
Graduates				
(2) Industry	8 (22)	20 (49)	10 (56)	38
Government	3 (8)	4 (10)	2 (11)	9
Abroad	9 (24)	5 (12)	4 (22)	18
Teaching	1 (3)	1 (2)	1 (6)	3
Higher Education	16 (43)	11 (27)	1 (6)	28
n (100%) =	37	41	18	96‡

Comparing industry to higher education:
(1) $x^2 = 8.16$; $df = 2$; $p = .02$.   (2) $x^2 = 11.55$; $df = 2$; $p = .01$.

   * after obtaining a higher degree.
   † 40 students working outside research/teaching excluded; these are classified as organisational scientists because of their occupation preferences.
   ‡ 3 students excluded for similar reasoning.

have them. I also would like to work with students, especially in the close way that a lecturer works with his PhD students. Finally, the salary is *good enough* [italics ours] for me.'

For some, the choice of higher education was rather a reflection of their dislike of the alternatives. For example, one graduate student said he wanted to work in a university because industry lacked any real interest in science. Some of his friends, he said, were 'frying sausages to determine the fat content'. Some saw the question in moral terms, saying, for example, that they would not work in industry in order to provide shareholders with dividends. Most of these students stated a preference for work in government laboratories. For example, another graduate student said:

'My opinions of industry are low and I therefore feel inclined towards working either in government laboratories or in a government-sponsored laboratory, e.g. Medical Research Council. I feel scientific research should not be directed towards putting money into shareholders' pockets, but should be directed towards curing disease, finding new ways of growing food and many other topics of pressing importance to humanity. Whereas I realise a government laboratory may be far removed from this ideal, I do not believe they are as far removed as industry.'

Another student, this time a graduate, classified as a private scientist, expressed a similar view:

'I have possible leanings to work in a government laboratory mainly because of the things which my friends have told me about various establishments. There appears to be more freedom of work and less "money-grabbing" tendencies in the Civil Service. I also abhor the thought of working specifically for one board of directors and their shareholders. Government laboratories seem to me to be a sort of midway type of work between the academic and the industrial.'

Again some students preferring to work in government made favourable comparisons between it and industrial employment. Thus one undergraduate, this time a public scientist, said:

'I have worked for two years in a government laboratory and also attended (part time) a technical college and hence have an insight into the working of both. My impressions of industry come only from interviews from which it seems apparent that: (i) one's research projects might be changed at any time because of higher incentives in some other field—rather frustrating for a scientist; (ii) opportunities for the "pure scientist" were less than for the non-academically minded person, although this is possibly changing slowly; (iii) with a BSc only, one would be expected to move to non-academic and non-scientific work within a few years.'

The general themes raised by students preferring to work in industry stresses the practical nature of the work which they defined as preferable to the pure research of a university, whilst others indicated that they had lost any interest in science.

Thus one undergraduate who wanted to work in industry and move quickly into management said:

'I chose to do a degree course in chemistry because it was the subject I could understand most easily, but I chose to come to university because it was both the logical next step after "A" levels and because it seemed to be an interesting thing to do. Now I do not want a permanent job in a laboratory as university life has completely killed all interest which I may have had in the subject—there is, I think, too much emphasis on laboratory work. Perhaps this is a good thing as it gives people like me a better idea of what they don't want to do.'

Another undergraduate with a similar career ambitions says:

'The promotion chances within industry appear to be more favourable, if only for the reasons that (a) there are more senior posts of responsibility, (b) more money available to pay salaries commensurate with responsibility.'

And, already looking forward to the future, he continues,

'My brief contact with industrial research laboratories indicates that there are many that could produce better results at lower financial expenditures and I would like to have a go at rectifying the situation during my career.'

A graduate student, an organisational scientist, illustrating his disillusionment with science says:

'Research has proved troublesome and results elusive. I am left wondering, in fact, whether I am suited to research, despite my former zeal. Industry would have the advantage of openings to other posts (i.e. other than the research/development) which could be seen and investigated while maintaining a research post. Furthermore, industry often takes greater pains to "settle" an employee into the right post than either universities or government establishments.'

A graduate, classified as an organisational scientist, gives his reason for wanting to work in industry:

'I worked in the Civil Service for six months, and there seemed to be too much red tape. After two years research I am fed up with pure science and fundamental chemistry and want to work on something more applicable to everyday living.'

Still another graduate student illustrates his organisational orientation towards the scientist's role, and the influence which this has on his preference for industrial employment:

'As a young man, I am considering marriage within the next two to four years, and I think industry with its higher rate of pay will give me more money to set up a home. There is also the consideration that some firms will advance money for house purchase at a much lower rate of interest than other sources.'

A private scientist illustrates nicely the balance between interest in science and its relevance for practical applicants:

'I should wish to follow up this year by trying for a higher degree in the pure sphere of university. For the most part I think that I should prefer work on projects, which might one day be applied practically, to working in the rarified university research atmosphere. This is why I hope eventually to work in industry.'

The reasons why some students wanted to go abroad again reflected their perception of the greater rewards which they expected to find. Thus one graduate public scientist says he has had:

'three years exasperation with poor equipment, no technical staff at all to assist in research (even to wash up). Is it any wonder that I rate a position in USA. They have good equipment and technicians. Red tape is too extreme in England, even for small purchases, and computer facilities are minimal.'

However, another graduate going abroad, this time an organisational scientist, does not criticise conditions in this country, and even suggests returning here, as do more than 60 per cent of graduates who go abroad immediately after leaving university, after a few years have elapsed. This particular student, who has been offered a post-doctoral fellowship in Canada, says his reasons for accepting it are, in order of importance: '(i) it offers an unrepeatable opportunity to go abroad; (ii) they pay £2,000 tax free; and (iii) I can work in university type research.' However—

'I anticipate returning in two years and continuing research in an industrial laboratory. I prefer industrial laboratories because I appreciate their "material" objectives—these are more rewarding than "academic" papers. Further, industry pays better than universities and government laboratories.'

Another organisational scientist going abroad also indicates how relevant it can be to a career in industry in this country:

'People trying for a job in industry in England as research chemists, find that top research jobs go to those with experience as a research fellow, particularly in America. I should like to gain this experience and yet receive a good wage for doing so; this is only obtainable in America. I shall only stay in America for two years, returning to industry in England afterwards.'

Although the association between career choice and scientific identity is significant, there are deviants however. For example, of the fifty-six public scientists among the third year under-graduates, twenty-three (39 per cent) said they intended working in industry. Similarly, eight of the public scientists among the graduate students (22 per cent) intended working in industry. There are also deviant cases with the organisational scientists, although far fewer of them. 13 per cent of the undergraduates and 6 per cent of the graduates who are organisational scientists prefer to work in higher education.

PREFERENCE AND OPPORTUNITY

We have suggested that in making their choices, individuals will seek those openings which they expect to be most rewarding. But they will also be influenced by their estimate of their chances of getting the desired job. For students, expectations about future academic results are clearly important in assessing their chances of an academic career. Thus we can predict that some of the 'deviant' public scientists are those who do not expect to do well in their degree examinations. Our evidence fully supports this hypothesis. Not one of the public scientists, who expected to obtain an upper-second honours degree or better, intended to work in industry after graduation, and only 17 per cent intended to do so after obtaining a higher degree. By contrast, 41 per cent of those who expected to do less well academically intended to work in industry after graduation. Furthermore, as many as 62 per cent of the former students hoped to work in a university, compared with only 10 per cent of the latter (Table 4.4). A similar picture emerges when we relate the career intentions of private scientists to expected degree result. The relation between scientific identity, academic achievement and career intention can be further illustrated by comparing the job preferences of those public and private scientists who expected to get good degrees. Whereas 62 per cent of the private scientists preferred industrial employment, only 17 per cent of the public scientists expressed a similar preference. Thus one undergraduate classified as a private scientist, said:

'I have no desire to get caught up in the academic rat race. I want to settle down to do research and not spend my time getting research grants.'

We are still left with the problem of explaining why some organisational scientists stated a preference for higher education. Such employment can, of course, be sought for its extrinsic rewards rather than from any intrinsic satisfactions from the

TABLE 4.4

Public scientists choosing a career in industry
do not expect to get good honours*

Occupational choice	Expected degree	
	First/upper Per cent	Other Per cent
Industry now	0	41
Industry later	17	26
Teaching	7	26
Government lab.	14	—
University	62	7
	100	100
	(n = 29)	(n = 27)

* Third-year undergraduates.

pursuit of knowledge. This is nicely illustrated by a quotation from one such undergraduate and may well be typical of the kinds of reasons why 19 per cent of the organisational scientists expecting good degrees intend to seek academic employment:

'I want to work in a university because it is a natural continuation. Also laziness because I don't care much for the biased pressures alleged to be applied in industry. Vacations and the lack of responsibility as regards producing a financially sound organic synthesis, for example, are pushing me to think deeply about university employment. I would like to teach but think that school teaching would be such a waste of training and also a waste of time as the salary is so low.'

He concludes:

'In any case, if one looks to university first for further employment, industry will always be there, if one is not bright enough or the salary is too low.'

If we take account of expected academic performance therefore, we are able to improve our prediction about career intentions considerably. Whereas 36 per cent of the public scientists stated a preference for academic employment, this figure rises to 62 per cent for those who expect to get good degrees and drops to only 10 per cent for those who do not (Table 4.4).

DIFFERENTIAL PERCEPTION AND CHOICE

But we are still left with the problem of explaining why some public scientists who expect to do well in their examinations still prefer industrial employment. A possible explanation is that they differ in their perception of industrial conditions. Our data lends some support to this interpretation. Such scientists have, in fact, a more favourable picture of the opportunities for publication, salaries, welfare and social benefits, and career opportunities for scientists in industry. Nor do they have such a favourable view of conditions in the academic world as those who intend to work there (Table 4.5).

We have sought to explain career preferences as a function of three main variables; scientific identity, perception and evaluation of job conditions, and expected academic achievement. We have been able to account for a large amount of the variance by these three factors, as hypothesised at the beginning of the chapter. The relation between the three variables is summarised in Table 4.6. As an example of the residual factors which may operate, we quote from a private scientist working for a PhD:

'I have been employed in industry since leaving school aged sixteen. By attending part-time courses I eventually obtained the qualification GradRIC and the ARIC. My firm (name

TABLE 4.5

*Public scientists preferring industry have a
more favourable image of conditions there*

Job condition	Image of adequacy* of provision in					
	Industry			University		
			Preferred employment			
	Ind. $n = 31$	Univ. $n = 36$	Per cent diff.	Ind.	Univ.	Per cent diff.
Salaries	84	61	+23[1]	27	25	+ 2
Holidays	43	37	+ 5	76	57	+19
Freedom to publish	33	14	+19	93	100	− 7
Freedom to choose research projects	10	8	+ 2	63	89	−26[2]
Technical equipment	81	82	− 1	60	59	+ 1
Social/welfare facilities	65	52	+13	47	73	−26[3]
Freedom to choose work colleagues	3	8	− 5	38	42	− 4
Career ladder for scientists	33	20	+13	52	56	− 3

* Percentages represent those saying provision of item 'more than adequate'.
[1] $x^2 = 4 \cdot 73$; $df = 1$; $p = \cdot 05$.   [2] $x^2 = 5 \cdot 26$; $df = 1$; $p = \cdot 05$.
[3] $x^2 = 5 \cdot 02$; $df = 1$; $p = \cdot 05$.
N.B. Actual number in each cell slightly differs—no response not included.
(Third-year and post-graduate students.)

mentioned) have since then sponsored me whilst attending
university to obtain the degree of PhD, I therefore feel obliged
to return to. . . . However, I would very much like to work in a
university research unit, in spite of the poor facilities regarding
equipment and apparatus, because I find that the atmosphere
is conducive to producing my best work.'

TABLE 4.6

Type of scientist, perception of conditions in university and industrial laboratories, expected degree result, and future employment preference

Type of scientist	Favourable employment conditions	Expected Degree result	*Future employment preference** University	Industry	$n =$ (100%)
Public					
	Better professional freedom in	High†	80	20	20
	university	Low	7	93	15
	Professional freedoms more or	High	67	33	3
	less equal	Low	20	80	5
			46	54	43
Private					
	Better professional freedom in	High	26	74	19
	university	Low	17	83	23
	Professional freedoms more or	High	22	78	11
	less equal	Low	30	70	7
			23	77	60
Instrumental					
	Extrinsic rewards more or less	High	25	75	4
	equal	Low	25	75	4
	Extrinsic rewards	High	22	78	11
	better in industry	Low	12	88	8
			18	82	27

* This does not include 10 students who preferred to work in government laboratories, 37 who preferred to teach, and 40 students who preferred to work outside the range of research/development/teaching.

† High means First or Upper Second, Low means any other response.

WOMEN AND SCIENCE

We have stressed the significance of scientific identity as a major variable. That is to say, we have argued that the kind of scientist that the undergraduate identified himself with will be significant. But we occupy a number of roles simultaneously. Every scientist is also a citizen, and possibly a husband and father. When he is acting as a scientist, he carries with him the 'latent' role of citizen. Some roles are more salient than others. For the dedicated scientist, his identity as a scientist takes precedence over his other latent identities. This may go some way to account for the fact that fewer women become dedicated scientists. They are not faced with the same crisis of identity which we have argued leads some to embrace the role of scientist. Moreover, they are aware of the latent role of wife and mother and of its incompatibilities with the role of public scientist. They will seek a role therefore, congruent with their dual identity. It is not surprising that 46 per cent of the women students intended to teach—significantly greater than would be expected[10] (Table 4.7). This is illustrated by one female student

TABLE 4.7

Women prefer teaching to research and development
(graduates and third-year undergraduates)

	Women	Men
Teaching	17 (46)	23 (9)
Research and Development, etc.	20 (54)	256 (91)
	37 (100)	279 (100)

who said:

'I have chosen teaching as a career partly because I feel that it is a challenging job when well carried out and partly because I am happy in an academic environment. I have spent three periods of working in industry in the summer vacations and the effect has been to put me off industry rather than attract me to it. If I thought I was of PhD calibre, I would probably try to remain in university, but unfortunately I have to admit to myself that this is not the case and this is generally borne out by examination results. When I complete my degree course, I am going to take the Diploma in Education and eventually teach in secondary grammar or comprehensive schools. Teaching, of course, is not entirely vocational. It offers a lifetime of security (provided one doesn't rob a bank), and the salary, whilst not competing with the fantastic incomes we hear about from our industrial friends, is quite adequate to maintain a respectable standard of living, and I feel that the salary is bound to increase in any case very shortly. A further advantage is that schools exist in every part of the country, so one can practically choose where one would like to live. Finally, there is the general steadiness of the job. One's time of corresponds exactly with the family's time off, so it is possible to lead a full home life with no difficulties about holidays, shift-work, and so on, to disrupt events, and this is by no means an inconsiderable factor.'

IMPLICATIONS FOR SCIENTIFIC MANPOWER

What then, of the problems with which we began this discussion, and in particular the view of the Committee on Scientific Manpower and the Swann Report that the universities encourage students to regard higher education and research as markedly preferable? Perhaps the universities do encourage this view—but they certainly don't succeed in persuading all undergraduates to adopt it. Indeed, as we have seen, many prefer industry. It is one thing to prefer a job in a university or industry and another to get it. Our figures however, suggest that the match between aspiration and achievement is not likely to be too far out. For example, 24 per cent of our third-year sample were intending to work in industry, compared with the 27 per cent who were first employed in industry for the period 1962–5 (Table 4.8). And 18 per cent of third-year undergraduates and 46 per cent of graduates preferred universities, compared with the 18 per cent who actually obtained employment there (1958–63).

TABLE 4.8

Occupation preferences of a sample of chemistry students

| Occupational preferences | Type of student | | | |
	First year Per cent	Third year Per cent	Graduate Per cent	All Per cent
Industry now	74 (30)	52 (24)	— —	126 (22)
Industry later	38 (15)	39 (18)	38 (38)	115 (20)
University	66 (26)	39 (18)	46 (46)	151 (27)
Other	73 (29)	87 (40)	15 (15)	175 (31)
	251	217	99	567

In short, by no means all undergraduates are persuaded of the superiority of pure research, and by and large, those who are not prefer jobs in industry. Thus far the fears of recent reports would seem to be exaggerated. But on some issues they are right. Our data supports the view that it is the most 'able'[11] students whom industry fails to attract. What is not absolutely clear is the exact nature of the association between expected class of degree, commitment to academic science, and choice of career. It would seem, however, that it is the attraction of an academic career which is of overwhelming importance. It is this which plays some part in the degree of commitment to science. And it is this in turn which is a major factor leading individuals to prefer a university to industry. But in addition to identification with public science, both a more favourable perception of conditions in industry plus expected class of degree below good honours influence the choice of industry (Table 4.6).

This argument also gains support from the data on the decision to pursue a higher degree. Public scientists were more likely to go on to a higher degree. But those who do not expect to get good honours, still intended to go into industry. Moreover, among graduate students, identity was still a major factor influencing choice; 56 per cent of the organisational scientists, compared with 22 per cent of the public scientists, intended to go into industry. Rudd and Hatch found similarly, that it was interest in the subject plus the enjoyment of research which was a main reason for going on to a PhD. Few did so specifically for instrumental reasons, such as a desire to become a university teacher. In other words, it is commitment to science which leads to graduate study. It is not the PhD as such which seduces many from industry who would otherwise be attracted. Indeed, many seem to stay on precisely because they have no career commitment; scientists most frequently gave reasons which came into the category of 'inertia'.[12]

There are two possible solutions. One is for universities and colleges to modify undergraduate courses to make careers in applied research more attractive, by placing less stress on academic values and greater emphasis on preparation for 'professional' roles, in which the application of knowledge is stressed. Sandwich courses in the technological universities and the substantial volume of undergraduate work leading to degrees of the Council for National Academic Awards (mainly sandwich courses) are examples of developments in this direction. Similarly, proposals have been made for an alternative type of PhD course to the present three years on original research which is heavily oriented to an academic career. The Committee on Scientific Manpower, for example, argued:

'There are grounds for believing that the present system of postgraduate support caters primarily for those who have already demonstrated high analytical abilities, and that it encourages able young people generally to embark on a career with a bias for pure research with its atmosphere of painstaking contribution to knowledge dissociated from economic necessities or limited timetables. We believe that in addition to the well established procedures for postgraduate support, leading primarily to the PhD degree, means must be found for providing support for those who evince high abilities more directly relevant to employment in industry.'[13]

This means that the SRC, the technological universities and similar institutions should modify the type of student selected for further education beyond the undergraduate level, and that courses should stress applied rather than 'pure' studies which dominate the traditional course.[14] This takes us back in part to the subject of this chapter and the previous one. In the light of the discussion here, it might appear at first sight as though the attempt to make 'applied' men out of those who are attracted to 'pure' science would fail. And for many, it undoubtedly would. However, the evidence also suggests that there is *some* room for manoeuvre. In the first place, the university *does* have some influence. Some universities produce a much smaller proportion of dedicated scientists (and it is the ratio with which we are here concerned). Secondly, the problem is not simply one of 'pure' research unsullied by practical applications. Research may be fundamental in the sense that it contributes to filling crucial gaps in theory, but at the same time have practical implications. Such work may yield even more satisfaction than work whose value is problematic. There may be many who would make the very small shift away from 'pure' research to fundamental work of this kind. And there is little doubt that universities who developed a tradition for such an approach would attract able scientists and produce dedicated men more amenable to some kinds of applied work.

It must after all be remembered that the scholasticism which considers any studies of practical value to be inferior characterises Europe and Britain rather than America. It probably owes as much to deeply-rooted aristocratic traditions as it does to the imperatives of science.

A second possible approach would be to consider the possibility of making careers in industry more attractive to the undergraduate. One solution put to us in our discussions with industry would be to change the image of industry. It is argued that the student picks up a faulty and distorted picture of industrial conditions from the university. Unfortunately, many of the conditions which are most disliked by the committed scientist do, in fact, exist. At least, this is the conclusion which we reach later in this book. And it is supported by the evidence from Kelsall's survey quoted in the Swann Report. In this situation it may not be so much a matter of modifying the present image as of modifying the conditions of employment so as to make them more attractive to well qualified graduates. Industry, however, may well fear that to modify these conditions may hinder the attainment of other, mainly economic goals, in which they are primarily interested. Thus by increasing the freedom of scientists over their research, and by allowing more freedom to publish, industry might find that too many scientists were becoming involved with status and recognition in the scientific community and as a result, commercial status and recognition would be demoted into second place. If this occurred, science might flourish, but industrial returns would decline.

Whatever the solution in detail (and more positive recommendations will emerge in Chapters 5 and 6), we are convinced that insufficient attention has been paid to this aspect of the problem. The Swann Report, for example, concentrates almost exclusively on recommendations for changes in education. While not disagreeing with the need for such changes, they cannot by themselves solve the problem. The fact remains that British industry needs to give much more thought to the way in which it uses scientists if it is to attract the good honours graduate who is, in general, in a position to choose.

NOTES

1. We are particularly indebted to Julienne Ford for her contribution to this chapter.

2. For a more detailed discussion of the theories and literature on occupational choice see J. Ford and S. Box, 'Sociological Theory & Occupational Choice,' *Soc. Rev.*, XV (1967), pp. 287–99, and J. R. Butler, *Occupational Choice*, H.M.S.O., (1968).

3. It is not, of course, being argued that all career choice is as rational as the example we are exploring here. See F. E. Katz and H. W. Martin, 'Career Choice Processes,' *Soc. Forces*, LI (1962), pp. 149–54. For a further discussion on this process of seeking to reconcile value and expectations, see E. Ginzberg *et al.*, *Occupational Choice: An Approach to a General Theory* (1951); Blau, op. cit., pp. 531–43; M. Rosenburg, *Occupations and Values* (1957); D. V. Tiedeman and R. T. O'Hara, *Career Development: Choice and Adjustment* (1963); B. S. Phillips, Expected Value Deprivation and Occupational Preference, *Sociometry*, XXVII (1964), pp. 151–60; A. Pavalko *et al.*, 'Vocational Choice as a Focus of the Identity Search,' *Brit. J. Counselling Psychol.*, XIII (1966), pp. 89–92. Other studies give rise, though less directly, to the same implications. Thus, for example, F. G. Caro and C. T. Philblad, 'Aspirations and Expectations: A Re-examination of the Bases for Social Class Differences in the Occupational Orientations of Male High School Students,' *Sociol. & Soc. Research*, LIX (1965),

pp. 465–75. They argue that class differentials in occupational choice derive not from different conceptions of the occupations but differing assessments of the chances of achieving them.

4. These propositions were formulated by G. Homans and J. Ford.

5. For details of the inquiry, questionnaire, and the sample, drawn from three English universities, see Appendix 3.

6. For a recent factual survey of scientists' salaries, see *Science Journal*, May 1965, pp. 81–5.

7. op. cit., Cmnd. 3760 (1968), p. 37.

8. For a forceful statement of this point of view and the implications for sociology, see H. Blumer, 'Society as Symbolic Interaction' in A. M. Rose (ed.), *Human Behaviour and Social Processes* (1962).

9. Our sample compares closely with a national sample of under-graduates and graduates; see Committee, op. cit. (1966), p. 11.

10. For an overall discussion of women in the professions, see J. A. Mattfeld and C. G. Van Aken, *Women and the Scientific Professions* (1967). See also P. Frithiof, 'Women in Science,' *Technology and Society*, Vol. 4, No. 4, p. 15–23.

11. It is, of course, problematic whether a high class of degree necessarily indicates ability. The little evidence there is does not show much relationship between university qualification and subsequent success. And the relationship which does exist may be more in the nature of a self-fulfilling prophecy. See, 'The relationship between college grades and adult achievement: a review of the literature,' *A.C.T. Research Reports No. 7* (1965), quoted in S. Hatch and E. Rudd, *Graduate Study and After* (1969).

12. S. Hatch and E. Rudd, op. cit. (1969).

13. Committee on Manpower Resources for Science and Technology, *Report on the 1965 Triennial Manpower Survey of Engineers, Technologists, Scientists and Technical Supporting Staff*, London, H.M.S.O., Cmnd. 3103, pp. 32–3.

14. ibid.

Chapter 5

Laboratory Scientists: Strains and Satisfactions

Our point of departure for this study was the fact that an increasing number of scientists work in industrial laboratories. The literature reviewed in Chapter 1 lead us to believe that such scientists experienced a variety of strains and dissatisfactions. Such strains stem in part from the fact that any kind of organisation imposes constraints and restrictions on the activities of the individuals who work in them. But for scientists organisational constraints are likely to be especially damaging. The bureaucratisation of science, it was argued, threatens the autonomy of the scientist, restricts his opportunities for the realisation of his skills and capacities and the exercise of creativity, and in extreme cases results in the alienation of the scientist—the loss of control over both the processes and products of his intellectual efforts. What, then, is the evidence for research and development laboratories in England?

SATISFACTIONS AND DISSATISFACTIONS

In order to discover to what extent scientists were satisfied or dissatisfied with conditions in industrial research laboratories, we asked our sample of chemists to rate each of twenty-one aspects of their jobs on a five-point scale from very satisfied to very dissatisfied. The distribution is given in Table 5.1 below. It is notable that there are relatively few aspects of their work which are found to be satisfactory by a large proportion of scientists. Only items 4, 7, 10, 11, 12, 17, 19 and 20 are reported to be satisfactory by more than 50 per cent of the sample.

On the other hand, at first sight the evidence hardly lends strong support to the more extreme arguments about the

TABLE 5.1

Sources of satisfaction and dissatisfaction among chemists in research and development

	Satisfied		Dissatisfied	
	Per cent	rank	Per cent	rank
1. Salaries	41	17	42	2
2. Free time	44	14	24	8
3. Supporting personnel	38	19	47	1
4. Research and development funds	59	5	20	16
5. Career prospects in science	37	21	38	3
6. Say in choice of projects	51	9	24	8
7. Conditions	67	3	19	18
8. Control over hours	48	12	20	16
9. Project termination	49	10	21	14
10. Complaints and suggestions	53	8	24	8
11. Publications policy	57	6	17	19
12. Patents policy	70	2	6	21
13. Prestige of laboratory	44	14	21	14
14. Rewards for productive workers	42	16	36	5
15. Staff-management relations	49	10	24	8
16. Long-term planning	38	19	38	3
17. Meetings and conferences	66	4	24	8
18. Recruitment policy	41	17	31	7
19. Organisation of research and development	55	7	22	13
20. Supervision	73	1	10	20
21. Use of skills	45	13	34	6

bureaucratisation of science. There *is* dissatisfaction over lack of autonomy (items 6, 8, 9) but these are ranked 8th, 16th and 14th and involve at most one quarter of the sample. In short, there is relatively weak support for the items in Kornhauser's list of potential sources of conflict and dissatisfaction. But there is much more support for Drucker's view that industry under-employs professional capacities. Over one-third were dissatisfied with the use of skills and capacities, rewards for productive researchers, the quality and quantity of supporting personnel, and the prospects for promotion as scientists. In short, our findings are in line with the evidence from other inquiries that, for many, industrial research offers only restricted opportunities for the challenging use of skills.[1] The capacities of many scientists are under-employed—a fact that is reflected in the major single source of complaint, the lack of adequate supporting personnel. As a consequence, the highly qualified are employed on relatively routine and undemanding work.

It is not surprising that the under-employment of skills and capacities is seen as a major source of dissatisfaction among scientists and is in line with what we could expect from the more recent work of Maslow and Herzberg. Although Herzberg's[2] findings are not directly comparable, his distinction between sources of dissatisfaction and satisfaction is worth discussion at this point. In his studies, Herzberg asked scientists and engineers to recall any time when they had felt exceptionally good about their jobs and also about events which had resulted in negative feelings about their jobs. He found that the events which were recorded as sources of dissatisfaction differed from the sources of satisfaction. The satisfiers were characteristically aspects of the work itself, such as a sense of achievement, recognition, and responsibility. The dissatisfiers were aspects of the context of the work as distinct from its content. They included items such as company policy and administration, supervision, salaries and working conditions. Such 'hygiene' factors were unlikely to be sources of positive satisfaction which could only derive from the work itself. By contrast, it is the job content factors which are the satisfiers.

Herzberg further linked his studies with Maslow's notion of a hierarchy of needs. At the base of the pyramid, there are basic biological needs, and at the apex, the need for what Maslow calls self-actualisation—tasks which are satisfying to the self. 'A musician must make music, an artist must paint, a poet must write, if he is to be ultimately happy. What a man can be, he must be. This need we call self-actualisation.'[3]

Now, if Herzberg is right, we could reasonably expect scientists in general, together with other professionals, to attach greater importance to self-actualisation. As a result of their substantial investment in education, their acquisition of a high level of skill and knowledge, they will look for jobs which enable them to exercise their skills and capacities. But we can also expect public and private scientists to place more importance on self-actualisation through scientific research, than do organisational scientists. We have defined the latter as those who are willing to use their science instrumentally, while the more committed scientist will attach importance to opportunities to *be* scientists—seeking an expressive rather than an instrumental role. And because they attach more importance to the actual nature of their work, they are more likely to be dissatisfied if this fails to meet their needs.

We can approach the question another way by drawing on the distinction between roles and identities discussed in Chapter 2. Dissatisfaction is more likely to arise when there is a lack of congruence between role and identity.[4] The scientist in the industrial research laboratory is expected first and foremost to be an employee, and to comply with the norms and expectations of that role in the economic system. Employers expect him, in return for his salary, to serve the company interests. And if the company interests require him to switch from one project to another or to refrain from publishing a paper of potential value to a competitor, then he is expected as an employee to comply. And for the scientist who identifies with the scientific community, and who accepts its norms and values the lack of congruence will be most marked. His identity as public scientist will conflict with the demands of his role as employee. And although he is identified as an employee by management (prescribed identity) he does not accept such an identification for himself.[5] But no such conflicts arise in the case of the organisational scientist for whom (as we shall see) there is little discrepancy between identity and role demands. Alternatively, the industrial scientist may stress his professional role, claiming autonomy to exercise his expertise. This is the identity which the 'private' scientist brings to his role. The view that the private (professional) and public (academic) identities are slightly improper is well reflected in the following remarks of research managers,[6] who frequently indicate where scientists of this kind should be employed.

'Man's first loyalty is to his family, then his firm, then to science. If he doesn't like his firm's attitude to research he should get out and take another job where his own ideas and those of his employer coincide more closely.'

'Scientists who cannot reconcile their conception of pure science with industrial research should transfer to academic establishments.'

ROLES, IDENTITIES, AND SATISFACTION

In order to explore the effects of identity on satisfaction, we used the classification of industrial chemists into the three types: public, private and organisational. Although we used value-dimensions similar to those employed on the students and modified our measure so as to make them suitable for the industrial situation, we encountered more difficulty in classifying industrial scientists than we had in classifying science students. That is to say, instead of finding only three clear types, we also found a substantial minority of scientists who did not fit into our framework. For instance, only 55 per cent of scientists who attached importance to freedom to publish also wanted high

TABLE 5.2

Types of scientist in industry

Freedom to publish important	Committed to a career in science	Desires autonomy at work now	$n =$	$\% =$	Classified
Yes	Yes	Yes	64	17	Public
Yes*	Yes/No	Yes/No	53	14	Deviant public
No	Yes	Yes	89	24	Private
No	Yes	No	36	10	Deviant private
No	No	Yes	92	25	Organisational (A)
No	No	No	38	10	Organisational (B)
			372	100	

* 'No' to only one.

levels of autonomy and were committed to a career in science. Our modified typology is shown in Table 5.2. It can be seen that the largest single group were organisational scientists (35 per cent), while only 17 per cent were public scientists, though we have classified another 14 per cent as 'deviant'[7] public scientists, on the grounds that although these attached importance to publication—the key characteristic—they attached importance to only one of the other two dimensions. Further confirmation that our measures were, in fact, distinguishing significant differences among our industrial scientists was found from questions probing reference groups—that is to say—those groups by whom the scientist wished to be thought well. The majority (54 per cent) of our two public scientist groups were oriented towards an external reference group composed of other scientists, compared with only 18 per cent of the organisational scientists.

The public scientists in our sample were in fact much more dissatisfied with the amount of influence research workers have in choosing projects (+30 per cent), supporting personnel (+25 per cent), ways in which projects are terminated (+25 per cent) and the over-all organisation of research (+20 per cent), (Table 5.3). In fact there was a markedly higher degree of dissatisfaction (10 per cent more) on ten out of the twenty-one items. It is particularly notable that items which appear as major sources of dissatisfaction for public scientists now include those which we would expect from the bureaucratisation of science theory—notably control over the choice and termination of projects.

These findings are consistent with what we had come to expect. Our criticism of much of the earlier work was that it had failed to take account of the differing extent to which graduates in science have internalised the norms of science. It is only those who are strongly committed to science (public, and to a lesser extent private scientists) who will experience major strains when they occupy roles which are not congruent with their identities. Kornhauser's work fails to bring out the implications of this distinction. It is only when we control for scientific identity that we find evidence of the kinds of dissatisfactions that can be expected to flow from the employment of scientists in bureaucratised settings.

TABLE 5.3

Public scientists are more dissatisfied than organisational scientists

	Type of Scientists		
	Public (+ + +) (*n* = 64)	Organisational[7] (– – –) (*n* = 38)	Percentage difference
1. Salaries	41	37	+ 4
2. Free time	25	16	+ 9
3. Supporting personnel	59	34	+25
4. Research and development funds	22	18	+ 4
5. Career prospects in science	42	37	+ 5
6. Say in choice of projects	36	6	+30
7. Conditions	25	13	+12
8. Control over hours	26	8	+18
9. Project termination	33	8	+25
10. Complaints and suggestions	25	24	+ 1
11. Publications policy	17	16	+ 1
12. Patents policy	8	3	+ 5
13. Prestige of laboratory	23	8	+15
14. Rewards for productive workers	33	34	– 1
15. Staff-management relations	31	26	+ 5
16. Long-term planning	42	16	+16
17. Meetings and conferences	25	21	+ 4
18. Recruitment policy	36	26	+10
19. Organisation of research and development	25	5	+20
20. Supervision	11	5	+ 6
21. Use of skills	28	24	+ 4

ROLE STRAIN

How serious are such dissatisfactions? To what extent do industrial scientists actually experience strain?[8] To explore this, we asked our respondents the following question:

'It is often suggested that there are conflicts between the values of science and the values of industry, and that scientists working in industry sometimes experience these conflicts in the form of difficulties and strains. Do you think some scientists in your research/development organisation including yourself experience difficulties? If so, could you give us a brief example of this?[9]

A substantial proportion (34 per cent) of all scientists reported experiencing difficulties personally. Moreover, such strains are more prevalent among public scientists (54 per cent compared with 28 per cent for organisational scientists—Table 5.4).

TABLE 5.4

Public scientists are more likely to experience strain than organisational scientists

	Public	Public deviant	Private	Private deviant	Organi-sational	
No strain	16 (25)	25 (47)	32 (36)	15 (42)	58 (45)	146
Strains experienced by other scientists	13 (21)	12 (23)	28 (31)	11 (31)	35 (27)	99
Strains experienced personally	35 (54)	16 (30)	29 (33)	10 (28)	37 (28)	127
n (100%) =	64	53	89	36	130	372

By far the largest single source of strain reported centred around the problem of autonomy, including dissatisfaction over the selection and termination of projects, the exclusion from the formulation of long-term planning, and the frustration at not being able to follow up ideas which were scientifically interesting.

'I have myself experienced something of the kind on the premature (from the scientific point of view) termination of an interesting project.'

'Termination of projects which are scientifically interesting but which are obviously commercially untenable has led to frustration. . . . '

'I think that frustrations are often caused by the fact that projects are frequently terminated when the lack of commercial success becomes apparent or when economic success is not apparent after a given expenditure, regardless of the novelty of the information that is being obtained.'

Others complain more specifically about the limitations on the pursuit of interesting projects and tight control over day-to-day working:

'I personally feel, along with many of my colleagues, that some of the work we are doing should be followed up in a more thorough and academic manner, but since this does not directly affect the firm, sanction cannot be obtained.'

'It is often impossible to follow up interesting, but probably unprofitable, lines of investigation because lines of investigation pursued must show a high probability of profitability to the company.'

'What is distasteful to me is the tight control over methods of attacking problems which management—for economic reasons presumably maintains.'

'Our section leader . . . takes a very great personal interest in everything you do which is extremely annoying, because anything you do you find him breathing down your neck wanting to know what you are doing and every little detail, when sometimes you just want to keep it to yourself until you see whether it's got any worth or not. . . . It's not so much the interest, because he interferes, he tells you how it should not be done, how you should set the apparatus up, how you would work it out, and this does get rather annoying after a while. We are intelligent people, and we are quite capable of working it out for ourselves.'

'As far as I'm concerned, there is too much (of supervision) and often it is not very good supervision; it means that you just don't have personal and mutual respect. He tends to get bogged down in minor details, experimental details which, as far as I'm concerned, are not his problem at all. He would be much better off putting more effort into co-ordinating the work of different people who work on the same project as himself.'

Closely related were a number of criticisms of the *quality of management*, and here a dominant complaint was that industrial management had little understanding of how scientists operate. In addition there were complaints about poor communication between scientists and managers:

'British industrial managements are not science-minded, that is, they are not able to make full use of their scientists because management does not know what their scientists can and cannot do. Research departments are not regarded as an integral part of an industrial organisation, but rather as something slightly embarrassing, which it is necessary to have but with which one cannot do very much.'

' . . . because many of the production departmental management are not scientifically trained and are therefore not fully appreciative of the values of science as opposed to their own immediate short term requirements in a problem.'

'There is not enough communication, nowhere near enough. I mean, the research director I've met about three times, I suppose, and these were very brief . . . other people I know share this feeling. The communication you get is rather channelled through the section leader. I know that these people are very busy, it would be nice if you're working in the lab. to know that somebody in higher management really is interested.'

'When sort of progress meetings occur on projects, or latest results are discussed or future policy on a project are discussed, although you are an actual chemist at the bench you are

100

not invited along. It may even be chopped up at divisional manager level. He goes to the meeting and he relays the result to the section leader and the section leader relays the results to you and the story's never complete all the way down; you only get the dribs and drabs, if anything at all. Half the time you don't even know whether the meeting has taken place, you are not really involved in anything very much."

A third area of strain, naturally more characteristic of public scientists, concerned communication with scientists working outside the organisation, through publication and the discussion of mutually interesting research ideas:

'There is some conflict on the question of commercial secrecy, and the wish to discuss problems with people outside on a casual basis.'

'This paper (to be part of a symposium), was whittled down by the requirements of industrial secrecy as interpreted by our management, so that it was not really worth presenting.'

'Publication policy is almost bound to seem arbitrary at times, but a particular example recently was the failure to give permission to publish a review article . . . '

'Permission to publish admittedly good work was refused on the grounds that it would show competitors a field of interest. This I find conflicts with a scientific attitude. Also older administrators have, on occasions, refused publication of work they could not understand. This arose from their desire for a very simple form of presentation of a scientifically advanced work. The resultant text-book treatment is thus refused in a first-class journal and only acceptable in a third-rate one which is seldom read.'

A fourth area concerned scientists' *ethical doubts*, expressed as comments that commercial interests were being given priority over the welfare of science and, by implication, humanity. Such comments reflect too, the strains generated by being both an employee and having an allegiance to the scientific community:

'This conflict is real since the aims of industry are mainly monetary, rather than the pursuit of knowledge for its own sake . . . occasionally this manifests itself as a feeling of contempt for the work.'

'It is an utter waste of time and human effort unravelling the composition of competitors' products, and endeavouring to 'fiddle' round patents. In an ideal society, this type of work would not be needed.'

Finally, there were numerous complaints about salary structures and career prospects. The latter we have reserved for a later section, but it might be useful here to indicate the typical complaints scientists made about salary scales, although it should be pointed out that we do not consider this complaint to be directly related to the discrepancy between the values of science and industry. Scientists' complaints about salaries do not appear to be directed at the level but at the mystery of assumed scales. For example, one scientist suggested:

'You see advertisements to employ chemists and they say the salary scale is higher than elsewhere. . . . when you join you're dead keen and work hard for your future, and after a few years you find that salary scales are things that exist only in the minds of the administration department, you are not told exactly what they are; if you complain, they say, "Oh, we do have salary scales," and if you ask what they are, they can't tell you, it's confidential. To all intents and purposes, they exist only in the minds of the administration department.'

A similar statement came from another scientist:

" . . . on the personal side, we are told that we get salary increments every year and at this time the section leader and the divisional manager together will tell you how you've done, how you could improve your performance and generally give an appraisal of the situation. Now this happened to me *about four years ago*; this year the section leader and the divisional manager came round and handed me an envelope in the corridor, and that was it.'

THE RESEARCH ENVIRONMENT

So far, we have explored the ways in which strains and satisfactions are related to the characteristics of the scientist. But just as scientists do not constitute a homogeneous group, so there are considerable differences in the research environments in which they work. Industrial scientists may be employed in long-term fundamental research which differs little from much of the research pursued in universities. At the other end of the research spectrum, there is relatively short-term development work and 'trouble-shooting', which may require them to switch their efforts immediately from less urgent work.

It is reasonable to argue that the public scientist would find working at the basic/applied end of the research spectrum more congenial, and would experience less strain and dissatisfaction than those public scientists employed in development/service work. We found in fact little difference in answer to the role-strain question. But when we looked at their satisfactions, we found that public scientists employed on basic/applied work were more dissatisfied with many aspects of work than were those in development/service work (Table 5.5).

Some clues to this situation can be found in the detailed comments made in reply to the role-strain question. One possible source of dissatisfaction with management and supervision springs from the anxiety felt by those on long-term work that their position is particularly vulnerable, as the following quotation suggests:

'Development predominates in my organisation and this is accentuated (on the instruction from top management, not research and development management) during times of financial strain on the company. This has not worried me, but colleagues on the more pure research side, I feel, object to being transferred to short-term development projects at such times.'

TABLE 5.5

Public scientists in basic/applied research
were more dissatisfied with many aspects of work
than those in development/service

Percentage dissatisfied with:	Basic/ applied research (*n* = 35)	Development service work (*n* = 29)	Difference	*p* =
Amount of free time allowed for private research	60	29	+31	·001
Inter-personnel relations between research staff and management	40	25	+15	·05
Long-term planning of research programme	49	25	+14	·01
Degree of supervision of your work	47	18	+29	·01
General physical surroundings	29	43	−14	·05
Recruitment policy	41	57	−16	·05

Another scientist, this one working in more basic research, complains about a similar event:

'. . . the effort directed in the direction of fundamental research has been recently cut to nothing, presumably for *short-sighted* [his italics] reasons.'

A similar story, not from a scientist in the same company:

'Although not engaged in pure research, our laboratory had for some years made valuable contributions to scientific knowledge. Under a recent reorganisation, our duties have been cut back to short-term research for purely industrial ends.'

It is also possible to argue that the worker on more long-term projects will experience a heightened sense of deprivation over any restrictions on publication. Such scientists are more likely to have research which is publishable, not only because the research is of the type more acceptable in academic journals, but also because the development worker's research is very likely to be more suitable for patenting. Thus the more long-term researcher may well feel doubly grieved at having work which is potentially publishable, but which cannot be shaped into a publishable form because of the lack of freetime, and the constant pressure to become engaged in the next commercially interesting project.[10]

The comparatively greater dissatisfaction with recruitment policy among public scientists currently engaged in short-term research also invites some explanation. From our interviews, the view emerged that some firms did not make it clear to their applicants that the type of work on which they would be mainly engaged would be of this type. Instead, their tasks were left vague, or hints were made that the applicant's research interest would be of concern to the company who would make efforts to see that these would be seriously considered. Thus the expressed dissatisfaction among this group of public scientists suggest that their expectations have been disappointed, and that recruitment policies should take care not to give a misleading impression about what the applicant can expect.[11]

But there is a further possible explanation of the greater dissatisfaction among those employed in basic/development work. As we have seen, industry recruits a much higher proportion of those scientists who are prepared to accept the role of employee; it is organisational scientists whom industry mainly attracts. But for the smaller proportion of *public* and *private* scientists, there is evidence that they make some adaptation to their role as employee. They are aware of the obligation of the role of employee. They recognise that in return for a salary and other benefits, that they have some kind of obligation to give something in return. Gouldner[12] argues that such a *norm* of reciprocity exists in all societies. It prescribes that individuals *should* feel a sense of obligation when they receive rewards from another. Evan[13] used his concept to account for the unexpected fact that scientists in applied work experienced more strain than those in development sections. This, he argued, was because they felt under a stronger obligation than those in basic research to bend their efforts to meet the needs of the firm, yet had more opportunity than those in development work to contribute to science.

We can only guess whether any such mechanism is operating among the scientists we investigated. It could be, for example, that those in development/service work accept their role and realise that it is unrealistic to expect free time, or the absence of close supervision. But those in basic/applied work may expect more freedom and consequently experience greater disappointment when they do not get it. Only further, more intensive research could confirm such speculations but comments from scientists in our sample provide some supportive evidence. One, for example, was quite clear that he was paid to apply science in ways useful to industry and must comply with his side of the bargain or get out:

'If such a conflict exists, the scientist should leave industry and go somewhere where there is a more academic leaning. Or he should modify his views—after all, to my way of thinking, an *industrial* [his italics] scientist is paid to apply science in a way which is useful to the industry—not necessarily of use to science. If the findings are of scientific use, this should be accidental rather than essential.'

Two other comments also illustrate the acceptance of the obligations to industry:

'Industrial research *must* [his italics] impose restriction on the scientist and he should be aware of this when he makes his decision to go into industry. Scientists today are still regarding themselves as members of an élite priesthood. They are not.'

'The conflict is occasionally noticeable in the case of scientists engaged upon basic research to supply knowledge to more applied scientists who are working on products. Certain of the basic people must be constantly reminded of their *duty* [our italics] in this respect.'

The fact that a scientist is employed in more long-term research does not then justify us in drawing any conclusions about the actual conditions he enjoys, such as autonomy and freedom to publish. We need a more precise measure of the organisational environment than a crude categorisation based on function. Nor can we conclude from observing that a substantial proportion of public scientists are concentrated in a particular research laboratory that the conditions are likely to simulate those in a university. It may well be, of course, that such a concentration generates a pervasive atmosphere[14] which helps to underpin their identities as public scientists, and hence to reinforce their claims to be treated as such.[15] However, it is important to verify this possibility and not simply infer its existence.

The following six aspects of the working environment are clearly important:

1. Do companies typically, allow scientists any free-time during the normal work week? This is important because, as we have already suggested, such time is essential for preparing work for academic publication.

2. Do they permit scientists any substantial influence over the selection and termination of research projects? This is important, not only because it satisfies the scientists' need for autonomy, but it is likely to increase the integration of the scientist in the organisation by allowing him to engage in research which is of mutual interest to both employer and employee.

3. Are scientists constantly brought into discussion with research management about the general social arrangements existing in the organisation, or does management keep an aloof distance from scientists thus giving the latter the impression that management regards them as employees to be instructed and not consulted?

4. Does the organisation's public policy result in a sufficiently high rate of publication, and are the research scientists included in the decision making process for categorising work as publishable? Naturally this condition is more relevant to public scientists than other types of scientist. As an organisational characteristic, publication policy assumes more importance, therefore, where such scientists are a substantial proportion of the total research population.

5. Is the type of work on which any individual scientist is engaged suitable in terms of his skills, capacities and identity, or are these considerations superseded by management's definition of priorities without regard to the matching of skills, identity, and work level?

6. Are scientists reasonably assured of their positions, not only in the narrow sense of there being no present threat of unemployment or redundancy, but in the more important sense of a guaranteed future? Do industrial companies, in other words, provide a career for scientists *qua* scientists, or do they expect scientists, for numerous and complex reasons, to relinquish research in preference for a non-research position? More briefly, to what extent does an industrial company simulate research conditions and a collegiate atmosphere typically associated with a university or fundamental research organisation?

In order to measure the degree of organisational freedom, we scored each company for each of the following variables:

(i) the percentage of scientists who were able either to select their own research projects or be influential in this procedure;

(ii) the percentage of scientists who had more than 20 per cent of their work time made available to pursue their own research interests;

(iii) the percentage of scientists who, either by themselves or in close contact with their immediate supervisor, were able to terminate research projects; and

(iv) the percentage of scientists engaged on basic or applied research;

(v) the percentage of scientists who claim they could publish more except for organisational hindrances.

The following table (5.6) provides these percentages for each company. Each company was then ranked for each organisational freedom, and the ranks were summated. We then dichotomised our sample into those laboratories with a relatively high rank for organisational freedom and those

TABLE 5.6

Company differences in the amount of organisational freedom and the proportion of public scientists

	Organisation							
	A	B	C	D	E	F	G	H
Percentage of time spent on basic/applied research	54[2]	27[4]	84[1]	21[8]	27[4]	35[3]	25[7]	27[4]
Percentage scientists free to select own projects/or be influential	46[2]	34[5]	73[1]	21[7]	43[3]	21[7]	43[3]	27[6]
Percentage scientists who can terminate projects, sometimes with assistance of immediate supervisor	35[4]	49[2]	73[1]	12[8]	42[3]	18[7]	25[5]	23[6]
Percentage scientists with more than 20 per cent free-time	12[5]	33[2]	38[1]	4[8]	13[4]	17[3]	6[7]	12[5]
Percentage scientists *not* hindered by organisation in their publication rate	86[1]	38[8]	45[7]	72[2]	48[5]	46[6]	63[3]	50[4]
Total rank score	14	21	11	33	19	26	25	25
Final rank	2	4	1	8	3	7	5	5
Classified High (H) Low (L)	H	H	H	L	H	L	L	L
Public scientists (Per cent)	35	18	39	28	34	43	24	35
Rank	3	8	2	6	5	1	7	4
Classified High (H) Low (L)	H	L	H	L	L	H	L	H

with less favourable conditions. In a similar way, we were able to divide our companies into two groups according to the proportion of public scientists they employed (Table 5.7).

TABLE 5.7

Typology of industrial research organisations

| | | Proportion of Public Scientists | |
		High	Low
Provision	High	1 (A,C)	2 (B,E)
organisational			
freedoms	Low	3 (F,H)	4 (D,G)

The actual distribution of companies between these four cells is shown by the letters in parenthesis (i.e. companies A and C fall into cell one).

We are arguing broadly that strains and satisfactions will be the result of an interaction between the roles which scientists are expected to carry out and the identities which they bring with them to the performance of such roles. Or more specifically that public scientists will experience more strains where the organisation imposes constraints on their performance as professionals and even more as scientists. We can, therefore, expect more dissatisfaction among scientists in cell 1 compared with cell 3. And because organisational constraints are likely to be a source of dissatisfaction for all highly qualified workers, we can also expect more dissatisfaction among those in cell 2 compared with cell 4, although we would not expect the differences to be so marked.

As can be seen in Table 5.8, column 1, those in cell 3 are in fact less satisfied, on no less than eleven items compared with cell 1. And those in cell 4 are more dissatisfied on nine items, but the differences are less marked. (Only one item is different a the ·01 level of significance.) In short organisational freedom is an important determinant of satisfaction for scientists, particularly in those companies which employ a high proportion of the more committed scientists.

SUMMARY AND DISCUSSION

In general then, we find relatively little support for the view that the employment of scientists in industry generates substantial and widespread strains. There *is* dissatisfaction over

110

TABLE 5.8

Satisfaction is related to both the proportion of public scientists and the degree of organisational freedom

Job item	Cell 1 cf. Cell 3	Cell 2 cf. Cell 4
1. Salaries of research/development chemists	†	*
2. Free time allowed	†	*
3. Supporting personel		
4. Amount and security of funds	*	†
5. Prospects for promotion up a scientific career ladder		*
6. Amount of influence in choosing research projects	*	*
7. General conditions of buildings	*	
8. Influence over hours of work	†	*
9. The way in which projects are terminated	*	
10. Attention by management to suggestions and complaints	*	
11. Publications policy		
12. Patents policy		
13. Prestige of the laboratory in the scientific world	*	
14. Way in which highly productive research workers are rewarded	†	*
15. Interpersonal relations between research staff and company management		
16. Long-term planning of research programme		
17. Opportunities to attend scientific meetings		
18. Recruitment policy for research staff		*
19. Overall organisation of research (team structure; dept. divisions)		
20. Degree of supervision over main work		
21. Extent to which capacities and skills are employed	*	

Number of items where there are significant differences between cells.

† Indicates difference highly significant at ·01 level.
* Indicates difference very significant at ·05 level.

lack of autonomy, but this only becomes marked for the more committed scientists. But we do find much more widespread evidence on the failure of industry to use the skills and capacities of its scientists to the full. And this is a particularly potent source of dissatisfaction for scientists, who like other professionals, have invested time and effort in acquiring qualifications.

The fact that many, indeed the majority, do not experience high levels of dissatisfaction and strain from working in industrial science is explained largely by the processes of differential socialisation and self-selection which we explored in Chapters 3 and 4. Industry does not in general attract the more committed scientists who would be most likely to experience the strains of bureaucratisation. And these firms which recruit the highest proportion are also likely to employ them more fully on basic/applied work rather than development/service work (Table 5.6).

But the evidence does not justify an optimistic or complacent conclusion. It is clear that the conditions under which scientists are employed, if they do not give rise to dissatisfaction, do not on the other hand generate high levels of satisfaction, with possible adverse consequences for motivation (Chapter 7). In particular, lack of autonomy and participation in many firms gives cause for disquiet. And the gloomy picture which many undergraduates have of the lack of opportunities for the full use of skills and capacities in many industrial laboratories is all too true.

The answer may be that many firms would do well to recruit more technicians, and to employ their more highly qualified scientists on more demanding work. But it may also be the case that British industry does not devote enough energy to more long-term and challenging fundamental projects which hold promise of practical application.[16] Such fundamental work must, of course, be followed through by development to production. But there is some evidence that such research strategies are relatively underdeveloped in Britain compared, for example, with the USA.

One conclusion of particular significance emerges from our data. There are marked differences in the administrative policies between laboratories. In company C, a high proportion of scientists have a say in the choice of projects on which the work, and the company ranks first on autonomy for project termination and free time for research (Table 5.5). True, its scientists also spend the greatest proportion of their time on basic/applied work. But company F which ranks third on basic/applied work and like company C employs a high proportion of public scientists, ranks seventh on project autonomy. There is clearly some room for manoeuvre. Although basic/applied laboratories in general grant more autonomy, not all do, and this is a matter which could well deserve the attention of research management. Or again, company F, which recruits the highest proportion of public scientists, yet employs only one-third of their time on more basic work, might well reconsider its recruitment and publications policies. Indeed, this company is particularly notable for the lack of congruence between the kinds of scientists it recruits and the way it employs them. The only item on which it is relatively liberal is in the amount of free time it allows for research not immediately related to company projects.

What then are the responses of scientists to the conditions which industry provides? One response, as we have seen, is that industry fails to attract most of the more highly qualified and committed scientists. But what of those who do choose a career in industry (or fail to find employment elsewhere)? What can they do, and what can management do? And what effect does industrial employment have on their motivation and productivity? We turn to these questions in the next two chapters.

NOTES

1. See in particular, the Swann Report: *The Flow into Employment of Scientists, Engineers and Technologists,* Cmnd. 3760 (1969); (Report of Kelsall's findings, p. 9); and V. Stanic and D. Pym, *Brains down the drain* (1968).
2. F. Herzberg, *Work and the Nature of Man* (1966).
3. Quoted in V. H. Vroom, *Work and Motivation* (1964).
4. The realisation of such harmony between the self and the roles it is expected to play is rare. Alter defines ego in ways which are favourable to alter in the sense that such definitions will carry with them expectations of role performance, beneficial for alter. It was probably this aspect of social interaction which led Sartre to conclude: 'Hell is other people.' Where a role is *embraced,* and is highly expressive of

the self and involves a high investment of self (in terms of time, energy, emotional resources), then any discrepancy between the self and a role will be a major source of strain.

5. For detailed examples of social labelling and attempts to avoid the consequences, see H. Patterson and E. Conrad, 'Shifting Sex Roles' in N. Johnston, *et al.*, *The Sociology of Punishment and Correction* (1962), pp. 140–4; B. A. Ward and G. G. Kassebaum, *Women's Prison* (1965). The consequences of social labelling are also being employed in the analysis of mental illness, see on this T. J. Scheff, *Being Mentally Ill* (1966). The central importance of social labelling and deviance is well argued in E. M. Lemert, *Human Deviance, Social Prospects and Social Control* (1967).

6. Research management may recognise the identity of a public scientist, but consider it inappropriate for industry—as in this quotation. Alternatively, we have some evidence that many research managers find it so difficult to conceive of the identity of public scientist, that they apply the identity of organisational scientist to all members of the laboratory staff.

7. We are, of course, using the adjective 'deviant' in the evaluatively neutral sense of deviating from the pure type. Table 5.3 refers to the extreme types 'public', and organisational (B), while Table 5.4 includes organisational types (A) and (B).

8. Role strain is being used in Goode's sense of the term: 'felt difficulty in fulfilling one's role obligations.' W. J. Goode, 'A Theory of Role Strain,' *A.S.R.* (1960), pp. 483–96. A. Southall, 'An Operational Theory of Role,' *Hum. Rel.*, VII (1959).

9. Evan employed both behavioural and attitudinal indicators of role strain. For the former, he used indices such as absence, late arrival, early departure, accidents. We have preferred purely attitudinal indices, on the grounds that behaviour is a function of both attitudes and situations. The presence of role strain cannot therefore be inferred simply from behavioural indices, such as absenteeism. W. M. Evan, 'Role Strain and the Norm of Reciprocity in Research Organisations,' *A.J.S.*, LXVIII (1962), pp. 246–54.

10. This is a further example of *relative deprivation*, first suggested by Stouffer as an explanation of a somewhat similar unexpected finding, that negroes in the American army were more satisfied than their conditions appeared to warrant—the clue being, of course, that relative to civilian life they were well off. See S. Stouffer, *The American Soldier* (1949).

11. For a more detailed discussion of this point, see G. F. Thomason, 'The Recruitment and Selection of Scientists,' in G. Walters and S. Cotgrove (eds.), *Scientists in British Industry* (1968).

12. The norm of reciprocity has a long history in sociological thought. It can be seen in Marx's notion of exploitation which refers to a breakdown in reciprocal functionality. The history of the concept since that time has been traced by A. W. Gouldner, 'Reciprocity and Autonomy in Functional Theory' in L. Gross (ed.), *Symposium in*

Sociological Theory (1959) and 'The Norm of Reciprocity,' *A.S.R.* (1960).

13. W. M. Evan, op. cit. (1962).

14. Such a subculture, related to the identities which individuals bring to their role performance can be conceptualised as a *latent culture*. Factors influencing the transformation of such latent cultures into manifest cultures include the relative numbers, physical proximity, communication between individuals and the perception that a collectivity or creation of a group will result in rewards or the avoidance of some of the sanctions of the formal structure. Thus, 'latent identities will not affect either individual behaviour within the group or the collective behaviour of the group unless they are in some way mobilised and brought into play in the daily interaction of group members'. H. Becker and B. Geer, "Notes on Latent Culture", *A.S.Q.*, 1960, 306–318.

15. In other words, as the proportion of public scientists increases, the resultant *structural effect* enhances their ability to protect their identities, and at the same time produces a more university-like atmosphere in the organisation. On structural effects see P. M. Blau, 'Structural Effects,' *A.S.R.*, XXV (1960), pp. 178–93, and for an exposition of a very similar concept 'compositional effects' see J. A. Davis, *Great Books and Small Groups* (1961).

16. Dr J. Goldman, reported in *The Times*, 18.4.69.

Chapter 6

Adjustment and Accommodation

We can no longer accept the simple picture presented in the earlier literature that scientists in industrial organisations experience strains resulting from the dissonance between the norms and values of science and those of industry. To speak of scientists or of research laboratories as though each conformed to some pure type, is, as we have seen, not empirically correct. Scientists differ; and so do laboratories. Moreover, the lack of congruence is modified by a selection process which lessens the possible dissonance.

But when all is said, some strains remain. And they need not necessarily be a source of anxiety, either from the point of view of the individual or the organisation. The view that organisations depend for their effective functioning on a substantial consensus over both values (goals) and norms (regulating means) has long since been challenged.[1] However, as there is some evidence in the next chapter that, beyond a certain point, some strains may hinder organisational effectiveness, we need to look at possible modes of accommodating strains. Firstly, what strategies are available to the individual scientist which will increase the congruence between his own needs and the demands of the organisation?

ROLE BARGAINING AND STRATEGIES OF INDEPENDENCE

Faced with a conflict between the needs of an individual and the demands of the organisation, there are a limited number of possibilities. The individual may seek to modify what the organisation demands. He may try to strike a better bargain by, for example, negotiating for more freedom to publish, or more money. That is to say, we can usefully think of the relation between individuals as a kind of transaction or exchange, in which each tries to maximise advantages and minimise costs.[2] Alternatively, the individual may devise a strategy which protects him from the demands of the organisation and enables him to assert a measure of independence or autonomy in the face of the demands of his employer.[3]

All organisations have to provide even the most menial jobs with some discretion simply because the cost of continuous observation would be prohibitive. Generally speaking, as the occupational prestige hierarchy is ascended, so the amount of discretion afforded increases. Although most of the studies of industrial strategies of independence have been confined to manual workers, it seems clear from our interviews that industrial scientists employ a number of strategies which enable them to pursue private interests whilst at the work place.[4]

One strategy is to formulate project plans for management's approval in such a way that materials can be acquired which are also essential for related research projects, which in their naked form would not gain management's approval. The following quotation illustrates both the notion of bargaining and the adoption of a strategem to increase autonomy.

'. . . we have a programme which is reviewed annually and I think that if the management thought that we were pursuing a fruitless line for any length of time, they would say so. But as a matter of fact this is not really likely to happen because in self defence we would not get ourselves into that position because we realise that as a rather long-term unit—well, I say realise, we suspect—that we are in a potentially dangerous position if there were a recession. People might say, "Well, you're not connected directly to operating processes, you're out first." We possibly have this at the back of our minds so when we are designing our programme, we do try to integrate the work we are going to do pretty thoroughly into what we think are the company's interest. . . . this is a defensive mechanism but we have so many lines of work which we would like to follow that we can generally choose work which is going to both be of interest to us and, we hope, useful to the company.'

A second strategy is to proceed with private research during those periods when formal commitments are not too intense or when surveillance is limited. This was illustrated by one scientist engaged mainly on long-term research:

'Occasionally we have thought there were one or two things we would like to follow up and we have been discouraged to some extent. Well, . . . we do have a tendency to do these *"under the bench"* so to speak, and one usually finds if anything promising turns up you can subsequently go to management and say this looks interesting and they will usually back you. One can do this to a limited extent; obviously you cannot go purchasing vast quantities of capital equipment without management's permission, but if there is something we think would be interesting and we have facilities already available for doing it, I am not averse to trying it out. . . . If I thought I might not get permission, it is quite probable that I would try it anyway.'

A third strategy involves exploiting the possibility of slight differences in approach consistent with achieving the main aim. A problem can be tackled in such a way that, if successful, the result will be patentable, but not readily publishable. Alternatively, the approach to the project can be such that, if successful, the result can be more readily transformed into a publishable form.[5]

One further strategy will increase the scientist's relative bargaining power. By sustaining a high rate of publication he can demonstrate his innovatory skills to management and at the same time, keep open the option of an alternative career in universities or research institutes. In this way, he may persuade his employer that it is in their interest to allow him all he wants in the way of freedoms and facilities. Although there is a limited willingness in industry to maintain scientists of exceptional quality in simulated university conditions, we certainly found a few examples in the firms we visited. These tended to be high-calibre scientists who had managed to satisfy both the above criteria.[6]

Before leaving this discussion of strategies of independence, mention should be made of the possibility of collective action.

Occupational associations, including both trade unions and professions, function to achieve betterment not only in the market situation, by improved incomes, security of tenure, and the like, but also in the work situation by regulating dangerous work, and protecting the skills of the worker against the threats from dilution. Although we were not able to extend our researches into this area, there is some evidence to suggest that lack of autonomy is related to collective action. Prandy's study of scientists and engineers[7] showed that those occupying relatively subordinate positions tended to support professional associations of the union type which stress collective bargaining. By contrast, those in the higher echelons belonged to professional associations which adopted strategies to raise the status of the occupation, and in this way to legitimise the high pay and dominant position of their members.[8]

ADMINISTRATIVE STRATEGIES

Naturally all these strategies depend partly upon the structural conditions of the work. Where supervision is close, and where scientists are not encouraged to initiate research programmes, the scope of such strategies will be less although not necessarily eradicated. Both previous work and the complaints voiced to us by scientists have suggested that supervising styles can have a significant effect on scientists' satisfactions and performance. Baumgartel[9] for instance, compared three styles of research leadership; *laissez-faire*, participatory and directive. He found in his sample of eighteen research laboratories in large American governmental medical research organisations, that the participatory style was positively related with researchers' sense of satisfaction with using their abilities, and with autonomy. From similar evidence, Pelz[10] has argued that a supervisory style which is collegiate rather than oppressive or absent, tends to be associated with more positive work attitudes and performances. Thus, where supervision tends to be strict, it appears that scientists are dissatisfied because this prevents any strategies of independence being effectively used. Where supervision is absent, it would seem that the total ambiguity of the situation generates feelings of normlessness which are also associated with low job satisfaction.[11]

The role of the research administrator is clearly of importance. It is he who mediates between the company and its scientists. He faces the difficult and challenging task of maintaining high levels of motivation and satisfaction among the scientists at the bench while at the same time ensuring that company objectives are achieved. How far is he aware of their needs? For without a sensitive awareness of the factors which influence their performance, he is hardly likely to make determined efforts to resolve the difficulties which scientists face.

TABLE 6.1

Comparison between the satisfactions of research administrators and research scientists

Item	*Percentage satisfied* Research admini- strators*	Research scientists†	Percen- tage differ- ence
1. Salaries	84	41	+43
2. Free-time	67	44	+23
3. Supporting personnel	52	38	+14
4. Research and development funds	68	59	+ 9
5. Career prospects in science	52	37	+15
6. Project autonomy	72	51	+21
7. Conditions	80	67	+13
8. Control over hours	78	48	+30
9. Termination of projects	64	49	+15
10. Complaints and suggestions	79	53	+26
11. Publications policy	88	57	+31
12. Patents policy	86	70	+16
13. Prestige of labour	57	44	+13
14. Rewards for productive workers	58	42	+16
15. Staff-management relations	66	49	+17
16. Long-term planning	45	38	+ 7
17. Meetings and conferences	99	66	+33
18. Recruitment policy	60	41	+19
19. Organisation of research and development	72	55	+17
20. Supervision	79	73	+ 6
21. Use of skills	55	45	+10

* Administrators $n = 69$
† Scientists $n = 403$

We asked research administrators to rate the list of job conditions for satisfaction, and also to say how satisfied they thought the chemists in their research and development sections would be with the same items. By contrast with the bench scientists, the overall level of satisfaction of the administrators was high (Table 6.1). But on many of the key items, the differences were less marked. On the termination of projects (+15 per cent), supporting personnel (+14 per cent) and the use of skills (+10 per cent), the satisfaction of the administrators was lower and the differences smaller. More important,

TABLE 6.2

Satisfaction of research administrators compared with their perception of satisfactions of research scientists

Item	*Percentage satisfied*		
	Research admini- strators*	Research scientists	Differ- ence
1. Salaries	76	29	+47
2. Free-time	48	35	+13
3. Supporting personnel	50	48	+ 2
4. Research and development funds	65	54	+11
5. Career prospects in science	49	19	+30
6. Project autonomy	67	27	+40
7. Conditions	78	66	+12
8. Control over hours	67	46	+21
9. Termination of projects	61	35	+26
10. Complaints and suggestions	71	43	+28
11. Publications policy	83	67	+16
12. Patents policy	81	72	+ 9
13. Prestige of labour	51	46	+ 5
14. Rewards for productive works	55	36	+19
15. Staff-management relations	64	33	+31
16. Long-term planning	43	30	+13
17. Meetings and conferences	95	68	+27
18. Recruitment policy	59	62	− 3
19. Organisation of research and development	70	46	+24
20. Supervision	84	55	+29
21. Use of skills	52	41	+11

* $n = 64$

121

however, is the administrators' *perception* of the levels of satisfaction of the bench scientists. It is an awareness of the problems which scientists face which is the prerequisite for effective action. The administrators as a group appear in fact to be well aware of lower level of satisfaction among the scientists they controlled (Table 6.2). Scientists were rated as less satisfied with career prospects (−30 per cent), project autonomy (−40 per cent), project termination (−25 per cent), complaints procedure (−28 per cent), publications policy (−16 per cent), rewards for productive workers (−19 per cent), staff-management relations (−31 per cent), attendance at scientific meetings (−27 per cent) and degree of supervision (−19 per cent).

However, an awareness of differences alone does not tell us how research administrators will respond. Indeed, it was clear to us, from many of the comments of the administrators on the strains experienced by scientists, that a sizeable proportion had scant sympathy with the dissatisfactions of the men at the bench. Indeed, a majority argued that if there were strains, then this was the fault of the universities. At least a quarter of the administrators said that the universities trained scientists to place too much emphasis on 'pure' science and basic theoretical research. But while we would not necessarily quarrel with this point of view, we would be concerned if it led to administrators dismissing any necessity to modify their strategies, the better to accommodate scientists, and to provide conditions for self-actualisation through participative leadership and opportunities for the full use of skills and capacities.

The response then of administrators to an awareness of the strains of industrial research is likely to be influenced more by the way in which they evaluate such problems rather than by any simple awareness of the fact that they exist. As a mediator between the needs of the company and those of the bench scientist, the administrator faces two ways. He is faced with the task of translating the predominantly financial goals of the company into meaningful and demanding tasks for the scientist. By top management, he is expected to guide his staff into research areas which are thought to be important to the company. By research workers, he has to be both an expressive leader protecting their professional needs and an instrumental leader by procuring adequate rewards and conditions. And like most 'men-in-between', he will himself be subject to the strains which such a position generates.[12] The closer in attitudes and position he is to top management, the more effectively he is likely to be in performing his instrumental role, but this may well be at the expense of his effectiveness as an expressive leader.

There is some evidence that being a good research man is a basis for approval and respect.[13] But there is also evidence that scientists prefer administrators who are seen as competent scientists and have technical competence in research. In other words, administrators who have a research background are more likely to be acceptable. Kornhauser came to the conclusion that the 'dominant pattern in industry is *not* to select research administrators on the basis of scientific competence . . . research administrators generally are selected on the basis of their capacity to fit into management'.[14] What then were the backgrounds and orientations of the administrators in our sample?[15]

All who answered this question were qualified scientists, 13 per cent by ARIC, the rest graduates with 44 per cent having higher degrees (34 per cent PhD or equivalent). But this alone does not tell us about their orientations. In order to probe this, we asked them about their career intentions (Qn. 13). 32 per cent would move into top management if they could, but 52 per cent intended to stay in research management. This information alone, however, tells us little. We also asked them what lead them into research management (Qn. 12). Few gave reasons which indicated a continuing interest in research. Only 16 per cent hoped to increase the representation of scientific interests to management by such a move, while another 22 per cent simply drifted and found themselves doing increasing administration. Such figures indicate therefore that the majority are likely to be company-oriented rather than science-orientated, a conclusion which is strengthened by their answers to the direct question on reference groups (Qn. 4). All of them answered this question in a way which indicated a company orientation: none gave scientists as their reference group. Similarly, on the questions on scientific identity, practically all were organisational scientists—only 6 per cent attaching importance to publication and 8 per cent considering a career in science to be important. Such findings are hardly surprising. As we show in Table 6.4, it is the organisational scientists who intend to switch to management.

From the data which we have, therefore, it seems very probable that many supervisors face one way rather than two, and that they are not likely to be very strongly motivated to pursue strategies which are aimed at tackling the strains and dissatisfactions which many of their colleagues at the bench are experiencing. Admittedly theirs *is* a difficult job. But the fact that only 17 per cent experience continuing strains *may* be evidence of their primarily company orientation and relative lack of sympathy with the viewpoint of the man at the grass roots of research. For it is only if they face two ways that they are likely to be able to function effectively as mediators between company objectives and the needs of scientists, and experience all the strains and difficulties which this involves.

SOCIALISATION

The needs and values which scientists bring with them into industry are the product of previous socialisation. But they are not necessarily fixed for all time. Organisations are aware of this,[16] and company induction policies are aimed at achieving some modification in attitudes. Moreover, attitudes are likely to undergo change simply as a result of experience which leads to their modification. Many new recruits to industry, fresh from the universities, are still imbued with the values of pure science. A number of scientists reported during interviews that although they had published when they first entered industry, they had since come to attach less importance to publication. It would be plausible to hypothesise that this re-socialisation process would result in the 'privatisation' of the public scientists, and a shift by the private scientist to instrumental involvement.

It was not possible for us to test this hypothesis in any rigorous way. We cannot, for example, compare different age cohorts of industrial scientists, since these may well have been changes in the recruitment pattern, with a higher proportion of public scientists entering industry in the late forties and a smaller proportion in the early sixties by which time there was a rapid expansion under way in the universities and colleges of technology. We were, however, able to test one hypothesis about the effect of socialisation. Abrahamson[17] has argued that although a basic disagreement about the appropriate level of autonomy exists initially between the scientist and his industrial employer, this disagreement disappears as the scientist lowers his estimate of the level of autonomy he considers appropriate, and as management increases the amount of autonomy with advances in seniority.

We tested this view by comparing the amount of autonomy desired with the amount received for scientists with varying lengths of industrial experience (Table 6.3). The general shift in the amount of autonomy granted to researchers tends to support Abrahamson's thesis that industrial management

TABLE 6.3

Both the amount of autonomy granted and the amount desired increase with length of experience

Degree of autonomy desired (D) and given (G)	Main type of research activity	Length of industrial experience			
		Up to 4 years percentage $n = 100$	4–9 years percentage $n = 100$	10–15 years percentage $n = 100$	16 years plus percentage $n = 100$
Project Selection (D)	Basic/applied	27 (41)	56 (41)	41 (29)	62 (29)
Project Selection (G)		4	15	24	31
	Percentage differential	−23	−41	−17	−31
Project Selection (D)	Development/service	27 (56)	41 (67)	42 (67)	39 (58)
Project Selection (G)		0	10	13	14
	Percentage differential	−27	−31	−29	−25
Project selection (D)	Basic/applied	65	76	82	90
Project influence (G)		35	39	55	70
		−30	−37	−27	−20
Project selection (D)	Development/service	74	68	77	74
Project influence (G)		14	29	43	45
		−60	−39	−44	−29

increases the freedom of the more experienced researchers. But instead of this narrowing the gap between autonomy desired and granted, it appears merely to keep pace with the expanding percentage of scientists who want high autonomy. For instance, we see that only a quarter of young industrial scientists wanted to determine their research goals, and that hardly any were allowed to do so. But of those scientists with more than fifteen years' industrial experience, nearly two-thirds wanted such a high degree of research autonomy, and one-half of these obtained it. The level of autonomy discrepancy was roughly the same. The same pattern of shifts over time occurred for scientists engaged on development or service work. If we define autonomy in broader terms to include the selection of goals and influence on those selecting goals, then the evidence does lend more support for Abrahamson's thesis, but only for those engaged in development/service work. In general then, for basic and applied industrial workers, it does appear that there is a constant level of autonomy discrepancy despite the increasing percentage of scientists who are given more autonomy.

This does not rule out the possibility of a measure of resocialisation. There may well be a decline in the importance attached to publication and in the desire to publish, masked to some extent by the tendency of the more instrumental scientists to leave to take up administrative posts. We found some evidence for this in our interviews.

CAREER MOBILITY

Where then there is dissonance between the needs of the scientist and the demands of industry, he may lessen the strains by protecting himself from the demands, by bargaining for more satisfying conditions, or by adjustment through moderating his expectations. In extreme cases, where such strategies do not produce a tolerable solution, he may leave and seek a more congenial environment. What are his chances of doing this?

We have already indicated that as a group, industrial scientists are less academically qualified than those employed in universities. The average industrial scientist cannot then view his chance of moving into the academic world with much optimism. However, even if the individual's qualifications are formally sufficient, there is always the problem of time spent in industrial employment. The longer a scientist has worked in industry the less his chances of moving into the academic world. This is not because there is any prejudice against industrial scientists, but simply that they tend on the whole to have a low standing in the scientific community because of their low publication rate. This does not mean that their work is necessarily inferior, but that it is less visible to the academic world compared with someone actually working in a university. Consequently, the university scientist's work is more widely known, and he enjoys higher prestige. This reputation is further enhanced by personal contacts and friendships.

But there are other possibilities of mobility. A move between industrial companies is both easier and more common. Thus, in our sample, 45 per cent had worked elsewhere (see Table 6.4). But of these, 58 per cent (104) had always worked in the industrial sector, and only 27 per cent (fifty) had previously worked in a university. However, movement between firms

TABLE 6.4

Previous employment of industrial scientists

	$n =$	Percentage $=$
Have always worked in the same company	223	55
Have always worked in industry	104	26
Have also worked in university	50	13
Have worked in government	31	8
Have worked outside industry abroad	17	4

Total $n = 403$; exceeds 100 per cent because some scientists have worked in more than other setting outside industry.

does not appear to be very frequent. For instance, of the 45 per cent who had worked outside their present company, as many as 62 per cent (seventy-nine) had, in fact, only worked in one other company, and a further 26 per cent (thirty-three) had worked in two other companies. From these figures it appears then that for most industrial scientists an industrial job is a once and for all time job, and that for a minority a second chance in another industrial position is about all that they can manage.

We would expect public scientists to be more mobile. It is they who are likely to have established a reputation and hence to have widened the market for their skills. Moreover, their loyalty is to the scientific community rather than to the firm. This relationship, between cosmopolitanism and mobility was suggested by Gouldner, and evidence has been presented in support of it by Caplow and McGee, Abrahamson, and Lewis,[18] but these studies have been concerned with university academics. So far as the authors are aware, there is no other study which has attempted to show that cosmopolitanism is related to mobility on a sample of industrial scientists.

We obtained two measures of mobility from our respondents. The first relates to their previous work-like history, and the second refers to their future employment intentions. Among our sample, we found that only 19 per cent had ever been employed outside the industrial complex, and of these, nearly four-tenths were public scientists. In addition, slightly more public scientists had been mobile between industrial firms compared with other types of scientists (see Table 6.5). When we look at scientists future career plans, it is among the public scientists that we find those who are considering a move outside industry, mainly to an academic post (Table 6.6). But the numbers are small. Only six public scientists were, in fact, planning to move to an academic post. And of these, five had been in industry for less than two years.

Why did these few public scientists intend leaving industry? One of them, a recently qualified PhD who had published three papers over the last two years, felt that he was considerably under-employed in his work. In addition, he complained about the limitation on publication, and the interruption of a research programme 'by urgent production problems'. Another public scientist who had had slightly more experience also claimed that he did not feel effectively employed and specifically complained about the termination of projects.

'From time to time', he said, 'a scientific problem crops up

128

TABLE 6.5

Percentages of public and other types of scientists
who have previously worked elsewhere

	Public scientists Per cent	All others Per cent	$n =$
Have worked outside industry	44	14	71
Have worked in other industrial firm(s)	31	23	90
Have always worked in same firm	25	63	211
$n = (100\%)$	64	308	372

which is of some academic interest to me and perhaps other colleagues, but which is not, or very soon ceases to be, relevant to the project concerned. We are then told to stop any investigations, even if they are quite trivial.'

A third public scientist, who had only two years' experience, complained bitterly about the absence of freedom at work, and was very critical of management. He considered that 'employers of scientists often do not know what they expect of

TABLE 6.6

Mobility intentions by type of scientist

	Public Pure Per cent	Public Deviant Per cent	Private Pure Per cent	Private Deviant Per cent	Organisational Per cent	
Research management	24	31	34	53	70	
Remain in research	63	65	60	45	26	
Go to another firm	3	1	6	0	3	
Go to university/ C.A.T.	8	2	0	0	0	
Other	2	1	0	3	1	
$n = (100\%)$	64	53	79	36	130	372

scientists and certainly give no clear directive on what they expect them to do'. The same scientist also criticised industry's publication practices. He suggested that:

'. . . withholding information causes wasteful duplication of effort and slows down progress enormously. I don't think that scientists are much influenced by "economic" stimuli (such as the knowledge that rival firms are working on similar projects) but the free interchange of knowledge and ideas, and the scientific rivalry which this engenders is invaluable for keeping enthusiasm alive. The tendency of some firms to publish results only in the form of patents is very bad. Not only are patents tedious to read, but they often contain red-herrings and allow no worthwhile discussion.' Clearly this scientist, who spent 80 per cent of his time on service work, perceives his work situation as extremely frustrating and since he has only recently left the academic world, doubtless feels that he has a good chance of returning.

Another PhD scientist with only two years' experience said that:

'. . . there is . . . a conflict between the desire to undertake more academic research pursued more for its intrinsic or scientific interest—and the need to conform to the work which is more commercially profitable for industry, and suffer. In my own case, I have found it possible to reach some compromise, fitting in a certain amount of fundamental research pursued in "free-time" with the provision of another service section. If this work continues to prosper, I may well be able to obtain some more suitable academic position within the next few years.'

It seems clear then, that only a very small minority of scientists contemplate moving out into the academic world, and these are almost certainly all public scientists, with only a few years' industrial experience, of which they are extremely critical. As for those scientists, who intend moving to another firm, only one had more than ten years' experience, and 90 per cent of the remainder had had less than four years in industrial employment. With this group also then, it appears that intended movement is largely confined to those who are at a comparatively early stage in their careers.

How far then, is job mobility a device for moderating role strain? We cannot assume that a man moves for this reason. He may take another post simply because it offers promotion prospects. The amount of mobility is relatively small. Moreover, it is mainly restricted to public scientists. But it is these who experience most strain. And it is for this group that we have evidence that particularly in the early stages of their industrial careers, these public scientists who are planning to move into the academic world, appear to be finding difficulty in adjusting to conditions in industry.

THE CAREER PROBLEMS OF THE MIDDLE-AGED SCIENTIST

We collected a good deal of evidence, both from scientists, and particularly from research management, which suggested that there are particular problems centring around the middle-aged scientist. Many scientists are somewhat apprehensive about their rather bleak prospects for further promotion, while research management is anxiously considering what it can do with the growing proportion of research staff who are now approaching middle-age, having entered as young men during the period of rapid post-war expansion. For management, the problem is what to do with men who, it is believed (rightly or wrongly),[19] are past their best as productive scientists, but do not have what it takes to hold down an administrative post.

The problems are particularly acute for the committed scientist, who wishes to continue in research. Quite apart from any question of declining creativity or productivity, he still faces the prospect in the majority of British firms of seeing other scientists, who switch to administration, overtaking him in salary and status. Here, industry contrasts with both the Civil Service and the universities, where a scientist as such can attain top positions.[20] But the alternative facing the more intrinsically involved scientist, namely switching to administration, is unlikely to appear attractive. The more he identifies himself with the world of science and accepts its values, the greater the strain that such a switch will involve. For him, the valued rewards are the intrinsic pleasures and satisfactions of research and publication, or recognition by his colleagues of his success in achieving scientific solutions to problems. The rewards of a career in administration are quite different, and are unlikely to be as highly valued. Of course, there is the possibility of some measure of resocialisation, of some transformation of identity. We have no evidence on the proportion who successfully make such a switch. We certainly gained the impression that many of the senior research management had, at one time, made their mark in research. What can be said with certainty is that only a minority of both public and private scientists were actively contemplating moving into administration, compared with nearly three-quarters of the organisational scientists (Table 6.6).

This leads us to a second reason for the higher dissatisfaction among the intrinsically involved middle-aged scientist. In our earlier discussion of the nature of science and of the scientific community, it was stressed that recognition is the reward for contributions to knowledge. In other words, a scientist is evaluated according to the assessment of his contributions by other scientists who are the only ones who are competent to judge. Thus the authority of the scientist among his colleagues rests on his reputation: it is a reflection of his expertise. In marked contrast, the authority of officials in organisations such as industrial laboratories, flows from their position in the authority structure. Compliance is achieved because, in the last analysis, officials can legitimately confer rewards and punishments.

This distinction between professional and administrative authority[21] raises problems for any organisation employing professionals. Universities minimise the problem by ensuring that administration is controlled by academics. Such considerations are also important in industrial research laboratories, especially for the more committed scientist, who is increasingly likely as he gets older, to find that his administrative superior may be an ex-research worker whom he perceives as a somewhat indifferent scientist.

But the scientist who wishes to remain in research, despite the limited rewards which it offers, faces other sources of anxiety. There is firstly the fear that he may be past the peak of his creativity.[22] There is certainly a widespread belief among research management that such an age decrement occurs. And it is not unreasonable to suggest that such a nagging fear may lurk in the mind of the scientist for whom research is of prime importance. Rather more important, however, is the tangible evidence of the rapid obsolence of scientific knowledge which is brought to the attention of the middle-aged scientist with every new recruit.[23]

What solutions are available then, either to the middle-aged scientist, or to management? For the scientist who does not wish to move out of research, or who does not possess the personal qualities necessary for a more administrative post, there are few alternatives. Some prefer to remain in work that interests them, and accept the career sacrifices which this involves. For others, the solution is to effect a measure of withdrawal from work and to invest in other areas of life with more significance.[24]

The strategies available to management to cope with the problem of the increasing proportion of middle-aged scientists are also limited. Transfers to senior positions in other departments raise fears about opposition from those already occupying positions of authority in those departments. Where the outsiders come from research sections which frequently have a tradition of political struggle with non-research departments, there may be even greater reluctance to stir up discontent by such moves. But in any case, it is doubtful whether there are enough non-research senior positions available for the potential candidates.

The problem of developing the necessary administrative skills is being tackled in some firms by sending such scientists on courses to colleges and universities, and to independent concerns such as the Administrative Science Trust. But the evidence we have suggests that the scale of the provision did not match the need. Moreover, doubts were expressed about the policy of selecting those who had already shown administrative potential, with the consequence that those whose need for training was greater were passed over.

For the more committed scientist, a more radical solution is required. The idea of a 'dual-ladder', with senior posts available in science as well as in administration has been much discussed and frequently advocated. But it has not, as yet, been widely adopted in British industry. It is important to examine reasons for this. The first reason seems to relate to the tendency of management to maintain the traditional distinction between 'staff' and 'line'. In this traditional conception, 'staff' are seen as operating only in an advisory capacity; they have been denied executive functions. Now the adoption of a dual career ladder, if carried to its logical conclusions, might eliminate the distinction between staff and line. For where high status can be attained within the staff, the status superiority of the line is threatened. The unwillingness of management to abolish traditional staff/line distinctions has been documented in a study of the British electronics industry.[25] In this study, it was found that the expansion of laboratory staff generated a number of political conflicts because scientists desired more power and higher status than they had hitherto enjoyed. But this causes the non-technical management to reassert the traditional staff/line distinction in order to avert the threat of scientists attaining increased executive power.

But promotion up a scientific career ladder is not without its problems for the scientist. The increase in freedom and discretion may be experienced by some as loneliness and a feeling of not belonging. The increased demands for the exercise of skills in interpersonal relations may be a source of strain for some, who, as we have seen, turn to science precisely because it deals with things rather than with people. Greater autonomy may also expose the scientist's limitations. Where there is little free-time, one can always rationalise any lack of creativity or productiveness. If the dual career ladder is to be a success, it is important therefore to stress the challenges and opportunities of such promotion, particularly the implications of added responsibility. The scientist must also face the fact that the increase in authority which should accompany his higher status is bound to increase his involvement in committees and paper work, which will take him away from the bench. It is essential too, to ensure that the scientific ladder does not become a solution to the problem of 'dead-wood'.

Any increase in opportunities for promotion may bring its own difficulties. The greater the opportunities, the greater the chance that those who do not get promoted become discontented.[27] There is also evidence to suggest that the routine system of promotion is preferable to the method of recommendation.

In the former system, the work of each scientist, as presented in the records, is periodically reviewed to determine whether it meets the standardised requirements for promotion. In the latter system, promotion only follows a personal recommendation from a superior. Thus the routine system represents an attempt to incorporate impersonal methods of control, and lessens possible fears of unfairness created by a situation in which personality as much as performance determines promotion. Furthermore, where requirements for promotion are standardised, the need to make one's work noticeable is obviated, thus reducing another source of anxiety for the less extraverted scientist.[28]

SUMMARY AND DISCUSSION

As we have seen, differential socialisation and occupational choice are major mechanisms which moderate the potential conflict between science and industry. And although it is the public scientists who experience the highest levels of strains and dissatisfaction, some remain for many organisational scientists. Scientists, like other employees, devise strategies of independence as a means of achieving some limited increase in autonomy. And some come to adjust by moderating their expectations. But dissatisfaction, particularly over autonomy, remains high.

The chances of escaping from unsatisfactory work by a change of job are limited, though this is the solution adopted by some public scientists after a few years in industry. Problems are particularly acute for the middle-aged scientist, having reached a ceiling in salary and status, facing anxiety about his continuing effectiveness as a researcher, yet reluctant or ill-equipped to move to administration. And although an alternative scientific career-ladder would mitigate the problem by providing senior scientific posts, it would not solve it.

There are limits, however, to the strategies available to the scientist to achieve greater satisfactions from his work. The main onus must inevitably rest on research management. And here the evidence does not justify any great optimism. Many administrators are unsympathetic to the very real problems of the bench scientist, preferring to place most of the blame on the universities. But however much one may agree with the view that industry is no place for 'academic' scientists, there remains the problem of devising as administrative procedure which would give opportunities for self-actualisation to highly qualified professionals by participative leadership and scope for the full use of skills and capacities through challenging work. There is a danger that the universities may be used as whipping boys to excuse administrative shortcomings.

One last consideration of key importance for the integration of scientists into the organisation is the position of research and development in the administrative structure. At the one extreme, there is the country-house laboratory, geographically and administratively isolated from production. At the other, there is the 'line-function' research and development section closely integrated into the administrative structure for production. Both have their problems. And the pattern which is appropriate for the more fundamental long-term research is not necessarily right for servicing and trouble-shooting. It is understandable that the strains and frustrations generated by too close an integration will encourage research and development sections to strive for maximum autonomy and separation.[24] But such a strategy may lead to increasing distance between the objectives of the researchers and those of the company. Yet for the more basic research, some separation is indicated, though the success of such separation will depend on the skill with which the research and development director can mediate research objectives to the laboratory. Too close an integration, with no buffer in the form of a senior scientist-administrator, can result in research and development becoming overwhelmed by the demands of production and servicing. In the larger companies, the obvious solution is some separation between sections concerned with production and those pursuing more long-term research objectives.

NOTES

1. On this, see D. Wrong, 'The Over-socialised View of Man' in N. J. Smelser and W. J. Smelser, *Personality and Social Systems* (1963). The point was also made forcefully by W. J. Goode who argues that strain is an inevitable result of the multiple role demands faced by all individuals, op. cit. (1960).

2. There is now a substantial literature developing the application of the notions of exchange, bargaining, and negotiation to the analysis of role relations and social systems. In addition to W. J. Goode, op. cit. (1960), see G. C. Homans, *Social Behaviour: Its Elementary Forms* (1961); P. M. Blau, *Exchange and Power in Social Life* (1964); and A. Strauss, *et al.*, 'The Hospital and Its Negotiated Order' in *The Hospital in Modern Society* (1963).

3. There has been a considerable literature on the concept of strategies of independence ever since Bendix made it famous with his much quoted elaboration of the concept, 'Beyond what commands can effect and supervision control, beyond what incentives can induce and penalties prevent, there exists an exercise of discretion important even in relatively menial jobs, which managers of economic enterprises seek to enlist for the achievement of managerial ends.' *Work and Authority in Industry* (1963), p. 251. This concept is closely linked theoretically with that of functional autonomy, since such strategies are only possible in a situation of functional autonomy. See, for an illustration, F. E. Katz, 'Explaining Informal Work Groups in Complex Organisations: The Case for Autonomy in Structure,' *A.S.Q.*, X (1965), pp. 204–23. In addition, see L. R. Sayles, *Behaviour in Industrial Work Groups* (1958); D. Roy, 'Quota Restrictions and Goldbricking in a Machine Shop', *A.J.S.*, LX (1954), pp. 255–66; and 'Banana Time', *Human Organisation* (1960); R. Blauner, *Alienation and Freedom* (1964), pp. 20, 43, 68, 73, 100, 104–5, 139. There have also been numerous studies on non-manual workers techniques of control: see for a summary of these C. Argyris, *Integrating the Individual and the Organisation* (1964), Ch. 5.

4. See for similar examples, Kornhauser, op. cit. (1961), pp. 64 ff.

5. This strategy was revealed to us by a scientist extremely interested in publishing for both intrinsic and extrinsic reasons.

6. The success of such strategies does not, of course, depend on the efforts of the individual scientist alone. His chances of influencing management may be increased, for example, if he is a member of a team which includes a high proportion of public scientists. This will facilitate the emergence of a latent-culture of professionalism which will both reinforce the values of the individual scientist and strengthen their bargaining position, since management will obviously be more reluctant to lose a group than one isolated individual.

7. K. Prandy, *Professional Employees—A Study of Scientists & Engineers* (1965).

8. For a discussion of the types of occupational association and their degree of 'unionateness' see: R. M. Blackburn, *Union Character and Social Class* (1967), pp. 18–43.

9. H. Baumgartel, 'Leadership Style as a Variable in Research Administration,' *A.S.Q.*, II (1957), pp. 344–60, and 'Leadership, Motivations and Attitudes in Research Laboratories,' *J. of Social Issues*, XII (1956), pp. 24–31.

137

10. For Pelz's most recent views on this, see D. C. Pelz and F. M. Andrews, *Scientists in Organisations* (1966).
11. Role ambiguity, defined as a different pattern of inadequacy in role sendings, has been shown to be related to job dissatisfaction and low self-confidence, and positively related with a sense of futility. See for evidence on this, R. L. Kahn, *et al.*, *Organisational Stress* (1964), esp. Ch. 5.
12. D. E. Wray, 'Marginal Men of Industry: The Foreman,' *A.J.S.*, LIV (January 1949), pp. 298–301.
13. N. Kaplan, 'The Role of Research Administrators,' *A.S.Q.*, IV, (1959).
14. W. Kornhauser, op. cit. (1962). Moore and Renck also found that industrial scientists had a generally unfavourable opinion of the technical competence of management. D. G. Moore and R. Renck, 'The Professional Employee in Industry,' *J. of Business*, XXXVIII (1955), pp. 58–66.
15. 22 per cent were heads or deputy heads of departments, 77 per cent were managers or section managers.
16. There is a considerable literature on socialisation in organisations which goes from the extreme views, such as W. H. Whyte, *Organisation Man* (1956) and C. Wright Mills, *White Collar* (1959), Ch. 10, in which the individual is viewed as being helpless to withstand the socialising influence of organisations, through to the more tempered writings which stress individual methods of coping, or working with the system. See on this E. Goffman, *Asylums* (1961). In addition, see, T. Caplow, *Principles of Organisations* (1964), Ch. 5, and S. Wheeler, 'The Structure of Formally Organised Socialisation Settings' in O. G. Brim and S. Wheeler, *Socialisation After Childhood* (1966).
17. M. Abrahamson, 'The Integration of Industrial Scientists,' *A.S.Q.*, II (1964), pp. 208–18. Marcson also makes the same point, see his *The Scientist in American Industry* (1960).
18. See Gouldner, op. cit. (1956); Caplow and McGee, op. cit. (1958); M. Abrahamson, 'Cosmopolitanism, Dependence-identification, and Geographical Mobility,' *A.S.Q.*, X (1965), pp. 98–106.
19. We examine this assumption in Chapter 7.
20. On this, see H. A. Shephard, 'The Dual Hierarchy in Research' in C. D. Orth, *et al.*, *Administering Research and Development* (1965).
21. For a discussion of this distinction, see P. M. Blau and W. R. Scott, *Formal Organisations* (1963). For a discussion of the implications of authority for scientists, see S. Marcson, op. cit.
22. This view that there is an age-decrement in creativity and productivity will be examined in Chapter 7.
23. The rate of growth of scientific knowledge is logarithmic, doubling approximately every ten years. This is another way of saying that half of what is now known has been discovered in the last ten years. D. J. de Solla Price, op. cit. (1963).
24. We have already made the point that the individual may invest varying amounts of his 'self' in his work roles. His latent identities as

husband, father, tennis-player, gardener, await development and he can balance his role budget, investing less time, interest, and energy in one role as he switches to another.

25. T. Burns and G. Stalker, op. cit. (1961).
26. It has also been argued that the idea of hierarchy is alien to the ideology of science and hence that the notion of a technical status ladder is inherently unacceptable. This argument, however, is not very convincing because hierarchies of status exist among academic scientists in universities and it is never suggested that these are inimical to the scientific ideology. See Orth, op. cit., Ch. 2.
27. This does not, of course, exhaust the possibilities. A partial solution for some scientists who fear the obsolescence of this expertise might be refresher courses. Others might switch to teaching in colleges of technology.
28. B. G. Glaser, *Organisational Scientists* (1964). His evidence is, however, confined to government sponsored medical research laboratories, and must be applied with caution to industrial laboratories.
29. T. Burns and G. W. Stalker, op. cit. (1961).

Chapter 7

The Productivity of Industrial Scientists

In previous chapters we have explored the interaction between the individual scientist and the roles which he plays. We have seen the problems which some face in articulating the sometimes conflicting demands of their roles in the social system of science and as employees in industrial laboratories. We have begun to explore the ways in which the interaction between their identities as scientists and the demands and expectations which stem from their occupancy of roles affects their behaviour. In this chapter we focus attention on one particular aspect of role performance—their output of various intellectual products including published papers and patents.

We face at the outset the problem of measuring productivity. Three broad approaches are possible. First, we could use colleague-evaluations of scientific performance. Secondly, we could ask scientists to make a subjective assessment of their degree of effectiveness in relation to their own standard of optimum effectiveness.[1] Thirdly, we can use objective criteria such as number of papers written, numbers published, numbers read at scientific conferences, and patents taken out. A number of variations on these alternatives are possible, for example, a count of significant contributions published in journals, which involves a judgement of importance.[2]

In this inquiry, the second and third of these criteria were used. Such measures are admittedly crude. They do not, for example, provide any indication of the quality of the product.[3] Above all, they do not distinguish between the 'creative' contribution, marked by a high degree of originality, and the run-of-the-mill routine publication.[4] But with these qualifications in mind, such criteria make it possible to throw some light

140

on the factors influencing the quantity of output, as a first step to a more refined analysis.

The interactionist approach adopted in this study suggests that the productivity of scientists will be a function of two main variables. Firstly, it is likely to be influenced by characteristics of the scientist, not only his knowledge and skills, but also his motivation to publish and patent. Secondly, the research environment can be expected to play some part. Basic research, for example, may offer more opportunities for work leading to publication than service work and trouble shooting. But we would also expect different environments to be more productive for some scientists than others. It is the interaction between a scientist and his environment which generates productive potential, and since neither scientists nor environments are homogeneous, we would expect that different types of 'mix' would produce different results.[5]

TYPES OF SCIENTIST AND PRODUCTIVITY

We may take it as axiomatic that an actor in any situation will behave in a way which he perceives as most likely to maximise his satisfactions and minimise his dissatisfactions. As we have seen in Chapter 6, this may involve a complex process of bargaining in which the actor minimises strains which result from the multiple and sometimes conflicting expectations of others in his role set. For the industrial research scientist, the context in which he functions is complex; and his definition of the systems of expectations will depend to some extent on his identity. The public scientist, for example, who identifies with the social system of science, will attach importance to publication as a means of achieving an adequate response through the acceptance of his product as a contribution to science. For him, a motivating situation will be one in which he can enjoy the reward of recognition for his gifts of knowledge by citations for published papers, prizes and honours. The more strongly he identifies with the world of public science, the more strongly he will desire such a response. We can hypothesise, therefore, that publication will be a function of strength of commitment to science.

But the industrial scientist is also an employee. As such, he is aware of the expectation that he contributes to company objectives by helping to develop and manufacture marketable products. In return, the company can offer a variety of rewards, such as increased salaries, titles, assistant staff, research funds. In order to receive such rewards, the scientist must pursue those goals most valued by industry by solving problems likely to bring commercial advantage. Such activities normally end with patenting the products of research. Since all scientists are in the same position as employees, such rewards are likely to appeal to all, although the public scientist will probably attach rather more importance to recognition than to economic rewards. We can hypothesise secondly, then, that the numbers of patents taken out will not be a function of degree of commitment to science. Indeed, we may find evidence for a weak inverse relation in those situations where effort devoted to patenting may hinder the achievement of publication by committed scientists.

Our findings confirm these hypotheses (Table 7.1). Controlling broadly for research environment which clearly affects the opportunity of carrying out research likely to lead to publication,

TABLE 7.1

Productivity and commitment to science
of chemists employed in basic/applied research

Percentage having in the last five years:	*Commitment**			$p =$
	High Per cent	Medium Per cent	Low Per cent	
Written one or more papers	46	32	21	0·05
Published papers	49	38	23	0·05
Taken out patents	54	56	53	n.s.
n (100% =)	35	37	43	

* High   = all three values important (i.e. autonomy, publication, commitment to a scientific career)
  Medium = two or more
  Low   = one or none.

we found that publication was significantly related to degree of commitment to science, but there was no such relation between commitment and patents.

RESEARCH ENVIRONMENT AND PUBLICATION

Productivity is a function, not only of the characteristics, of the scientist, but of the environment in which he operates, including the opportunities he has to work on projects leading to publishable papers, and the hindrances to publication resulting from the firm's publications' policy. And it is also plausible to argue that the scientist's effectiveness will be influenced by the more general organisational constraints that are to some extent inevitable wherever the activities of individuals must be co-ordinated and directed towards organisational goals.

It is reasonable to suppose that the industrial scientist will be influenced to some extent by the goals of the particular subsystem of the organisation in which he works. There will clearly be less opportunity in development and service work for basic research leading to the possibility of publishable papers. Moreover, such contributions from the scientist in development work are less likely to be valued by his employers and therefore less highly rewarded than contributions to product development. Table 7.2 confirms that such differences influence

TABLE 7.2

Research environment, commitment, and productivity

| | Commitment | | | |
| | High | | Low | |
	Basic/ applied	Dev/ service	Basic/ applied	Dev/ service
Index* of papers published	0·89	0·41	0·37	0·44
Index of patents	1·34	0·79	1·17	0·65

* Index: 5-year average of 0·25, score = 1
       5-year average of 0·75, score = 2
       5-year average of 1·25, score = 3.

productivity. Development and service work provides a less favourable environment for both publication and patents but, again, its adverse affect on publication does not appear to influence the lower output of those with little motivation to publish. In other words, the higher motivation of the 'committed scientist' to publish demands a favourable research environment, but this factor alone does not improve the contributions to science by publication from the non-committed scientist.

HINDRANCES TO PUBLICATION

Motivation to publish becomes a particularly significant factor influencing productivity in industrial laboratories, where scientists are likely to experience considerable hindrances to publication. It is well known that firms are generally anxious to see the findings of their research scientists fully covered by patents before they are published in journals or in other ways. A majority (56 per cent) of the scientists in our sample considered that their own publication was hindered, mainly by company policies. Restriction on publication was the most frequently given reason for failing to publish as much as possible (Table 7.3).

A major source of restriction is the company's desire to withhold information which may be of use to competitors, though some industrial scientists argued that this results in unnecessary restrictions:

'Industry withholds a great deal of information which could be published with benefit to all. Much is withheld through confused thinking about commercial security. It is always easier to say "No", than to make an objective appraisal and decision. A publication has become a record of a complete achievement. There is a wealth of incomplete and semi-empirical work—which will not measure up to the standards demanded in written publications—which could emerge in discussion at meetings and conferences or in the correspondence columns of the journals. Employers are not sympathetic to this kind of publication because they see no means of supervising it.'

'Yes, there most certainly is. In some cases there is the most

144

TABLE 7.3

Reasons given for believing that publication rate is lower than it could be

	n	Per cent
1. The organisation for which I work does not allow every research finding to be published	167	42
2. I write more than I publish; I am too critical and send very little off for publication	30	7
3. After a project is finished to the satisfaction of the company more work would be necessary to make it adequate for publication; I am not allowed that time during normal working hours	101	25
4. Most of my important findings are patented, and therefore the need to publish is diminished	91	23
5. My family life occupies the time I would need if I were to increase my publication rate	61	15
6. My other leisure pursuits do not leave me enough spare time to increase my publication rate	51	13
7. The company insists on complete coverage by patents before publication; by this time I am no longer interested	30	7
8. My skills are underemployed on work that is not publishable; if I had different, publishable work, my rate would increase	61	15

childish and unnecessary secrecy over work which is frequently of no use to competitors.'

Apart from restrictions in the interests of industrial secrecy, many (25 per cent) reported that insufficient time or resources were allowed to take the work to a stage at which it would be publishable.

'Work necessary after the initial success of any piece of research or development to bring it up to publication standard is not always sanctioned, nor the time. In many cases patent applications also hinder publication.'

'Yes, far too much new knowledge remains unpublished because of pressure of work and/or trouble encountered with obtaining permission to publish. On the other hand, much is published which would be best left unsaid.'

145

Some scientists, however, while recognising the existence of restrictions, considered them to be justifiable.

'There is a failure. However, this is understandable because of security reasons, information published must not be of use to competitors. The publication of knowledge by individuals in industry does not necessarily improve the scientific standing of the company, unlike the situation in the universities, and therefore there is no encouragement given to publish.'

'Of course. However, such knowledge will have been obtained because of investment by industry and industry can be under no obligation to reveal it; particularly as it would probably place their competitors in a more advantageous position (saving of money in duplicating unprofitable research, etc.).'

By contrast, others saw the restrictions as understandable, but wasteful, and necessitated only in a competitive economy.

'There is certainly a failure to publish knowledge, much of which would be of interest to other scientists. While our present political system remains, however, it is necessary to keep information secret; in the interests of the country this necessity results in much duplication of effort which I feel is wasteful.'

The way in which the scientist perceives the existence of any restrictions on publication depends very much on his own identity and the extent to which he has accepted the norms of science. The following statement is a forceful example of this from one who has clearly rejected the notion of the scientific community. He is not being held back from publishing: he is simply not motivated to publish.

'I do not think industry has any obligation towards other scientists in publishing information which is often necessary to maintain their competitive position *vis-à-vis* other companies. I do not subscribe to the view that scientific knowledge is something special or that we should recognise such an idealistic myth as an international "scientific brotherhood".'

146

Finally, a number of scientists argued that too much was published already. Apart from those who thought that much publication was trivial, there were some who favoured the face-to-face communication of 'invisible colleges' rather than the impersonal publications of the scientific community. Such scientists would be likely to feel more strongly about restrictions on attendance at conferences.

'There is too much publication—no one has time to read it all anyway. There should be much greater mobility of scientists from firm to firm and a free exchange of information on a personal basis at all levels.'

STRAINS, SATISFACTIONS, AND PRODUCTIVITY

We have already shown that industrial scientists experience strains and dissatisfactions, and that these are likely to be more marked among the more committed scientists. How far do such strains influence the effectiveness of the scientist?

At this point, we can usefully draw on a distinction made by Stein[6] between scientific and professional roles. The characteristic of the professional role is the importance attached to autonomy. The professional's authority is based on his expertise; on his specialised professional knowledge. On the basis of this expertise, he claims the right to exercise his independent judgement in carrying out his professional role. Thus, although the committed scientist can be expected to experience particular frustrations resulting from hindrances to publications, all scientists can be expected to demand adequate autonomy. In our sample, 75 per cent considered autonomy important, compared with 32 per cent attaching importance to publication. It is limitations on autonomy, therefore, rather than hindrances to publication that are likely to be prominent in the dissatisfactions of industrial scientists. And it is to lack of autonomy in particular that we can look for sources of strain and any possible influence this might have on productivity.

In Chapter 6, we demonstrated the extent of dissatisfaction with autonomy, and showed that in spite of increasing autonomy with age, the demand for autonomy also rises, with the result that the gap between autonomy granted and expected persists. Here, we are interested in the effect of this 'autonomy discrepancy' on productivity. We did, in fact, find significant differences in the productivity of scientists who were dissatisfied with the degree of autonomy which they experienced.[7] But this difference was confined mainly to the more committed scientists (Table 7.4). And it was related only to productivity as measured by publications, having no effect on patents (Table 7.5).

TABLE 7.4

Autonomy discrepancy reduces the productivity of the more committed scientists

Type of scientist	Autonomy discrepancy	*Productivity**		$n =$ 100%	$p =$ †
		High	Low		
Public	No	67	33	18	
	Yes	33	67	46	·02
Public deviant	No	29	71	21	
	Yes	19	81	32	n.s.
Private	No	43	57	47	
	Yes	27	73	52	·01
Private deviant	No	37	73	22	
	Yes	29	71	14	n.s.
Instrumental	No	20	80	59	
	Yes	17	83	71	n.s.

* One point was scored for each of having in the last five years (1) written a paper, (2) published a paper, (3) patented, (4) addressed a conference. High = a score of 4, 3, 2.
† x^2.

We also obtained a subjective measure of effectiveness by asking how effective scientists felt in their research, to compare their rate of research success with their estimation of the normal, and to estimate the proportion of their projects which are successful. Again, those who feel more highly productive experience less dissatisfaction with autonomy (Table 7.6).

Satisfactions with a wide range of other employment conditions were also explored. But although the committted scientists reported a significantly higher degree of dissatisfaction, we

TABLE 7.5

Autonomy discrepancy (public scientists) is related to
publication but not to patents

Autonomy discrepancy	Has published in last five years Per cent	Had work patented in last five years Per cent
No	72 (18)	56 (46)
Yes	28 (18)	48 (46)
	$p = \cdot01$	$p = $ n.s.

were not able to discover any clear connection between dissatisfaction and productivity. Dissatisfaction with autonomy, we have seen, *was* related. This came up again when we compared the productive (published in the last five years) with the non-productive. The non-productive were more dissatisfied over autonomy, project termination, use of skills, and supervision, but those who had published differed on other items, being more dissatisfied over funds, working conditions, hours

TABLE 7.6

Public scientists experience inadequate
autonomy feel less effective

Score on subjective productivity index	Autonomy discrepancy		
	No	Yes	$n = 100\%$
High 6,5	41	59	17
4	31	69	13
3	25	75	12
Low 2,1,0.	13	87	15
			57*

* Seven scientists omitted; no answers to these questions.

and prestige of laboratory (Table 7.7). The lesser dissatisfaction of productive workers over autonomy may reflect the fact that the more productive workers were given more autonomy. This is in line with the findings of Pelz and Andrews. Although 'well-adjusted' scientists produced more reports, 'work of greater scientific value (as well as greater usefulness to the

TABLE 7.7

Differences in dissatisfactions related to publication by committed scientists

Major dissatisfactions of those who have not published in last 5 years	Per cent Not published	Per cent Published	Per cent Differential
1. Autonomy over projects	63	41	+22
2. Termination of projects	65	25	+40
3. Supervision of work	44	24	+20
4. Use of skills	82	41	+41
5. Management's notice of complaints	56	47	+ 9
Major dissatisfactions of those who have published in last 5 years	Per cent Not published	Per cent Published	Per cent Differential
6. Security of funds	47	25	+22
7. General working conditions	29	6	+23
8. Autonomy over hours	59	25	+24
9. Prestige of laboratory	65	20	+45
10. Over research organisation	41	31	+10

organisation) was more likely to be done by scientists who did not fully see eye to eye with the organisation.'[8]

This lack of association between satisfaction and productivity may well arise out of the fact that any measure of satisfaction is, in fact, a measure of the discrepancy between a need and the extent to which it is met. High satisfaction may therefore indicate simply a weak need. Thus, those scientists whose need for self-actualisation in the sense of opportunity to make use of skills, and to work on difficult problems is high, may well be both more productive and more dissatisfied. When Pelz and Andrews distinguished between the desire for various conditions and their provision, they found that these were better predictors than satisfaction. And provision appeared to be quite strongly related to productivity, both provision of opportunities for self-actualisation and for advancement in the organisation.

These findings are in line with a substantial volume of evidence on industrial workers which suggests that happy and contented workers are not necessarily the most productive.[9] It is motivation rather than satisfaction which is important. Herzberg[10] used a similar distinction in his analysis of engineers and accountants, in which the motivators were factors related to the performance of the job itself, including achievement, growth and other aspects of what may be loosely defined as the opportunity for self-actualisation. Dissatisfactions were related to the context of the job and include pay and working conditions. A follow-up study by Scott Myers[11], which included scientists, identified achievement and the work itself as major motivators and company policy and administration as the main dissatisfiers.

Our own work is broadly in line with such findings. The less productive were most dissatisfied with 'motivators'—autonomy and use of skills, while the more productive expressed discontent with aspects of the context of their work, working conditions, funds, and the prestige of the laboratory (Table 7.7).

LABORATORIES COMPARED

In Chapter 5, we classified laboratories by two main clusters of variables—organisational freedom and the degree of commitment to science of the scientists employed. If research freedom is associated with research performance, then we would expect to find that scientists in companies in cell 1 of our typology are more productive than those in cell 2, and similarly those in cell 3 should be more productive than those in cell 4. Similarly, if commitment to science is an important variable, we would expect to find that scientists working for companies in cell 1 would be more productive than those in cell 3, and those in cell 2 should be more productive than those in cell 4.

In order to test these hypotheses, we employed both objective and subjective indicators of scientific performance. In addition to the objective measures already discussed, scientists were also asked how normal they regarded their research performance in comparison with other scientists engaged on similar work, and the responses to this were employed as our subjective measure of scientific productivity. We found that in those companies with a high proportion of dedicated scientists, organisational freedom make a considerable difference to subjective feelings about performance. Moreover, it was only in those companies with a high degree of organisational freedom that dedicated scientists felt more productive[12] (Table 7.8).

TABLE 7.8

Percentage of scientists who report work performance is above 'normal', related to organisational freedom and proportion of dedicated scientists

Proportion of dedicated scientists	Organisational freedom		
	High	Low	$p =$
High	60	35	·01
Low	43	42	n.s.
$p =$	·02	n.s.	

The publication of papers appears to be related to both variables. When commitment to science is controlled, a lack of organisational freedom reduces publication, although the influence is slight for those with a low degree of commitment. Controlling for organisational freedom, a decline in degree of commitment is similarly associated with a decline in publication. Of the two variables, 'dedication' to science would appear to be slightly more important than organisational freedom (Table 7.9). A similar attempt to relate patents to the two variables failed to find any significant differences (Table 7.10). It appears that differences in organisational freedom and the proportion of dedicated scientists influences only publications, that is, that type of productivity which is directed towards achieving recognition by the scientific community. Regardless of the percentage of dedicated scientists or the amount of organisation freedom, all companies have roughly the same proportion of scientists who have had work patented in the last five years. And it is this type of intellectual product which is most relevant

TABLE 7.9

Percentage of scientists who have published in last five years, related to organisational freedom and proportion of dedicated scientists

Proportion of dedicated scientists	Organisational freedom		
	High	Low	$p =$
High	46	28	·05
Low	24	10	·1
$p =$	·01	·01	

TABLE 7.10

Percentage of scientists who have registered patents, related to organisational freedom and proportion of dedicated scientists

Proportion of dedicated scientists	Organisational freedom		
	High	Low	$p =$
High	31	39	n.s.
Low	43	40	n.s.
$p =$	n.s.	n.s.	

to the work of the company. The only slight (but not significant) variation was found between those companies with a high proportion of dedicated scientists who have more organisational freedom, compared with all other companies. In the former two

153

companies, the rate of patenting was slightly lower than in the others,[13] thus confirming our earlier prediction.

AGE AND ACHIEVEMENT

Since the publication of Lehman's work[14] over a decade ago, there has been considerable controversy over the relation between age and achievement among scientists. Lehman argued that the creativity of the scientist reaches a peak typically in his late thirties and thereafter declines. Subsequent work has shown that a peak appears earlier in highly abstract disciplines, such as theoretical physics, and later for the more empirical disciplines such as biology. Furthermore, the peak is more pronounced for the most outstanding achievements, and flatter for minor contributions.[15]

Such findings have not gone unchallenged. They have been criticised by some as a statistical artefact.[16] Controversy apart, such an approach is clearly limited. The demonstration that there is a statistical association between age and achievement directs attention to the search for intervening variables. We cannot assume that any alleged decline in productivity is the direct result of some biological ageing process. Available evidence on the decline of intellectual abilities with age cannot account for either the age at which the peak is believed to occur in different disciplines, not any second subsequent minor peak. Moreover, it must be remembered that Lehman's work was concerned with creativity, whereas our interest here is with productivity.

This whole question is clearly of more than academic interest. The view that the productivity of research scientists begins to decline after the age of forty was mentioned to us by research administrators[17] as one of the reasons for transferring middle-aged scientists from research to administration. Industry is currently much preoccupied with this question. The rapid expansion of industrial research in the post-war period has resulted in a heavily skewed age-structure in industrial laboratories, with a concentration in the under-forties (Table 7.11). These men are now reaching an age when they are believed to be past their research peak. However, the generally accepted view that the expected decline in their productivity is due to individual rather than to environmental characteristics limits the strategies available largely to a transfer to some part of the organisation where such declining scientific creativity is of less importance—that is, out of the research laboratory.

TABLE 7.11

Age distribution of qualified scientists (chemists) in research and development

	Number of years in industrial employment					
0–3	4–9	10–15	16–21	22+		
105	111	96	40	51	=	403
25·8	27·6	24·1	9·8	12·5	=	100%

Recent researches have challenged the view that there is any necessary age decrement in productivity. Pelz has discovered a second peak some ten to fifteen years after the first. This peak Pelz found to occur five to ten years later in development-oriented laboratories than in research-oriented laboratories (i.e. in which research administrators attached importance to the publication of paper and to the advancement of scientific knowledge). In research, the first productive peak occurred between ages thirty-five to forty-four and in development between forty-five and forty-nine. Moreover, there was remarkable similarity between the laboratories whether they were located in industry, universities, or in government service. But in government laboratories, performance in development rose sooner and then declined, while in industry, performance continued to improve with age. And among PhDs in research, those in government laboratories were more susceptible to the erosion of age than those in universities.[18] On the other hand, the opposite trend was noticed for non-PhDs (assistant scientists).

Our own evidence also indicates an association between age and productivity, with an increase in papers and patents among those who have been in industry for more than twenty

155

years (Table 7.12). But such evidence needs to be interpreted with caution. Our data suggests that older and younger industrial scientists cannot be treated as a homogeneous group. Many scientists switch from research to administration at around age forty. Such a switch is particularly attractive to those who are less strongly committed to science as a career and who attach less importance to publication. We found, for

TABLE 7.12

Productivity and length of industrial experience

Percentage who have in the last five years:	0–4	5–9	10–15	16–20	21+
Published paper	29	25	26	18	48
Taken out patent	16	44	45	35	60

example, a higher proportion of PhDs among those with more than twenty years experience.[19] Moreover, the 'committed' scientists were more heavily concentrated in basic/applied research. It is this group, including a higher proportion of PhDs and strongly committed scientists, whose productivity is high and are being compared with a rather different type of younger scientist. Only by controlling for all the variables which may affect productivity can we allocate any residual influence to age alone, as distinct from (for example) motivation and opportunity to publish and patent. But such evidence certainly challenges the view that there is any necessary age decrement in productivity.

One further point. Pelz found age differences according to the objective measures of productivity used. Younger scientists produced more unpublished reports. Moreover, judgements of usefulness rose more steadily with age than did scientific contribution. But in their late fifties, judgements of usefulness declined rapidly for non-scientists in development laboratories despite the fact that their patents reached a maximum after age fifty. Pelz and Andrews consider it likely that the earlier peak represents work of a more 'divergent' or innovative type while the later peak is work of a more 'convergent' or integrative character.[20]

The weight of Pelz's and Andrews' evidence indicates that the major factor accounting for a decline with age is reduced motivation rather than any decline in intellectual powers. This is in line with the evidence already examined which attributes substantial importance to motivation as an explanation of differences in productivity. Pelz similarly found that strong involvement affected early distinction, and sustained achievement from erosion with advancing years. An analysis by Pelz of data obtained by Lieberman and Meltzer[21] found that those with little or no interest showed only one spurt at age forty and thereafter a decline. A study of Nobel laureates found that they began publishing earlier but also continued much longer; the nine who had passed the age of seventy were all still publishing.[22] Pelz and Andrews also found that older scientists were more likely to be productive if they had preserved a flexibility of approach, by avoiding over-specialisation in their thirties through having performed several different kinds of research and development activities.[23] Continued achievement also depends on a high level of self-confidence, evidenced by a willingness to take risks and to rely on one's own judgement.[24]

SUMMARY AND DISCUSSION

Two major factors affect the productivity of the scientist in industrial research. Firstly, his own characteristics, notably his specific hierarchy of needs, and secondly, the organisational context in which he works. These cannot be understood in isolation, since a motivating situation only exists when the environment offer rewards which are valued by an individual. Thus, the extent to which a situation helps an individual to maximise his effectiveness will be specific to the needs of the individual. In short, the prescriptions for the most effective use of instrumental scientists in development work are not necessarily the same as those for the public scientists in basic/applied research. Nor are the factors which maximise publications necessarily the same as those for optimum performance in the kind of research most likely to lead to patents. Thus, we found that both commitment to science and organisational freedom were important for publication, but neither factor was significantly associated with patents. Given a high degree of commitment to science, organisational freedom is positively related to publication, but in those companies with few dedicated scientists, organisational freedom makes little difference.

It is commitment to science, then, which is the necessary condition for a high level of publication. Only those who value the rewards which follow from recognition by the scientific community are motivated to publish. But this alone is not sufficient. The extra intensive effort required to transform research findings into a publishable paper is most likely within the context of a higher level of organisational freedom, including a more permissive attitude towards publication. By contrast, patent registration is valued by the company and more likely to be rewarded by them. It has some appeal, therefore, to all scientists, and, unlike publication, demands little further intensive effort beyond the solution of the company's research problems. Industrial scientists may then be 'local-cosmopolitans' in the sense that they seek rewards from both the organisation and the scientific community. This is an example of 'feudalism',[25] whereby the scientist is able to satisfy both organisational needs and his own needs for recognition. That is to say, the scientist contributes most of his product to the company, but he is also able to keep back some of it for his own use. This is only possible, however, where there is a measure of organisational freedom.[26]

The evidence on the significance of organisational freedom needs to be interpreted with caution. As we have seen, it is the discrepancy between the amount of autonomy desired and the amount experienced which is related to productivity (particularly publication). And although high organisational freedom contributes to higher publication rates and to a higher level of felt effectiveness among committed scientists, this tells us little about the precise level of autonomy which is optimal. There is a good deal of evidence[27] to indicate that some measure of control and co-ordination is more favourable to effectiveness than too little. That is, the completely autonomous scientist is not necessarily the most effective. Pelz and Andrews found that 'scientists performed better when some influence on their

158

important decisions was *shared* with several persons at various levels.'[28] Our measure of organisational freedom was concerned largely with the amount of influence scientists had over their work. It is influence rather than autonomy, therefore, which is related to effectiveness. In general, the optimum situation appears to be one in which the scientist has what Pelz calls 'controlled freedom'. In Pelz's words, a man 'was shown what mountain to climb and then it was up to him to get to the top.'[29] Other studies confirm that interaction with a research director who does not determine the procedures is more likely to result in innovation.[30] There are, however, differences between development laboratories and those mainly concerned with fundamental research. In the former, performance was highest when the goals were decided jointly by the scientist and his chief, but in research laboratories, the best performance was where 'self and colleagues' established goals.[31] Again, performance improved as autonomy increased from low to moderate, but it declined when individuals had more than 50 per cent say in setting goals—except for PhDs, where the peak was higher. In other words, the more highly qualified perform better with relatively more autonomy.

Our findings that satisfaction is not related to productivity does not mean that industry can afford to ignore the problem. The evidence certainly supports the view that the provision of conditions for both self-actualisation and advancement are related to productivity, even where the provision falls far short of what is demanded. Furthermore, it would seem that autonomy is more important for highly qualified personnel (PhDs), and more important for basic research than it is for development work. But our measures of productivity were somewhat crude. It may well be that the relatively smaller significance of both motivational and organisational factors on patents is misleading, since patents may be an unreliable indication of effectiveness in development and service work. Moreover, it must be remembered that dislike of the more restricted climate of industry leads many of the most able scientists to reject careers in industrial science, and others to leave industrial laboratories in the search for more congenial conditions. Thus, although attempts to meet the particular needs of scientists may possibly result in the less efficient organisation of industrial research, it may still be a strategy necessary to retain good scientists in industry.

Our researches, and the others we have quoted, also challenge the view that scientific contribution necessarily declines with age, and call for a re-evaluation of the policies in some firms which seek to transfer the middle-aged scientist out of research. The findings reviewed in this chapter reinforce the suggestions in the last that more could be done to sustain the productivity of the scientist at relatively little cost compared with the potential return. Possible action includes refresher courses, and positive policies to expose the scientist to a series of challenging and stimulating problems particularly in the middle and late thirties, as well as recognising the continued contribution of many by the introduction of a dual career ladder. It is noteworthy that the older scientists in our sample were particularly dissatisfied with the rewards available for productive scientists. Most are faced with a ceiling to their scientific career, which can only be raised by shifting from research to administration. Such conditions can only weaken the motivation to productivity.

From the limited evidence of this and other researches, there can be little doubt that the intellectual productivity of the scientist is greatly influenced by his research environment. Moreover, the fact that there are such substantial differences between companies indicates the urgent need for managements to become more aware of the effects which their actions have on the effectiveness of scientists.

NOTES

1. Studies by Pelz show a connection between subjective and objective measures indicating that the subjective judgements 'are not capricious, but based on objective performance', D. C. Pelz and F. M. Andrews, *Scientists in Organizations* (1966).
2. Such a measure comes closer to what is meant by 'creativity'. For a full discussion of creativity and its measurement, see C. W. Taylor and F. Barron, *Scientific Creativity: Its Recognition and Development* (1963).
3. A study of 120 university physicists found that the quantity of papers published was highly correlated with quality, although some physicists produce many papers of little significance and others produce a few papers of great significance. S. Cole and J. R. Cole, 'Scientific Output and Recognition; A Study in the Operation of the Reward System in Science,' *A.S.R.* (1957), pp. 377–90.

4. The terms 'creativity' and 'productivity' are used somewhat loosely in the literature. 'Creativity' has attracted particular attention from psychologists, who are concerned particularly with the novelty of the individual's contribution. Our investigation is into the 'productivity' of scientists, as measured by end-products such as published papers and patents, though it could, of course, be argued that creativity and productivity are related. See N. W. Storer, *The Social System of Science* (1966), Ch. 4.

5. This approach explains behaviour as a 'function of the transactional relationships between the individual and his environment. . . .' M. I. Stein, *On the role of the industrial research chemist and its relationship to the problem of creativity* (Univ. of Chicago—mimeographed).

6. M. I. Stein, op. cit.

7. Our data does not, therefore, entirely support Pelz's conclusions, since we found that productivity was related to the satisfaction of the scientist's demand for autonomy, rather than to the absolute amount he enjoyed.

8. op. cit. (1966), pp. 114–18. This measure of congruence was the extent to which the things a scientist enjoyed doing (contributing to a product with high commercial success; publishing a paper, etc.) were rewarded by advancement in the organisation.

9. J. H. Mullen, *Personality and Productivity in Management* (1966). For a review of twenty studies attempting to measure the relationship between job satisfaction and productivity, see Victor H. Vroom, *Work and Motivation* (1964), pp. 181–6. In general the evidence is in the predicted direction, although not significant. The evidence we have would tend to support R. Likert, *New Patterns of Management* (1961), in which he argued that the higher the skill required, the stronger the relationship between satisfaction and productivity.

   For a critical review of productivity and industrial research scientists, see J. R. Hinricks, 'Creativity among industrial scientific research,' *American Management Association; Management Bulletin* No. 12, N.Y. (1961).

10. F. Herzberg, *et al.*, *The Motivation to Work* (1959). See Chapter 5 for a more detailed discussion.

11. M. Scott Myers, 'Who are your motivated workers?' *Harvard Business Review* (Jan.–Feb., 1964).

12. We need to be cautious in the meaning we attach to statements by scientists about their productivity. For example, in Table 7.8, when scientists were reporting the normality of their work, they were probably using different criteria. Thus, the dedicated scientists may have in mind the norm for publications whereas other scientists may be referring to the successful completion of company projects and normal rates of patenting. This may explain why in Table 7.8 only the dedicated scientists differ in their reported rate of success according to their degree of organisational freedom. Organisational freedom only affects that form of productivity relevant to the scientific community; it does not affect the rate of patenting.

13. But these were also the companies with the highest percentage publishing. Such a situation may well generate tensions between industrial management and research scientists. If, when scientists are granted more organisational freedom, they respond by being more productive for the scientific community, but less productive for the organisation which employs them, then this may explain why some industrial companies do not grant more freedom. However, we should be very cautious before we accept this as the reason why research administrators behave as they do. Our evidence is at most only indicative. Furthermore, the relations between organisational freedom and productivity must be more fully explored before we can draw any policy conclusions.

14. H. C. Lehman, *Age and Achievement* (1953). 'The chemist's most creative years,' *Science*, CXXVII (May 1958), pp. 1213–22. 'The age decrement in outstanding scientific creativity,' *American Psychologist*, XV (Feb. 1960), pp. 128–34.

15. For a summary and discussion of this and other research, see D. C. Pelz and F. M. Andrews, op. cit. (1966).

16. See W. Dennis, 'The age decrement in outstanding scientific contributions: fact or artefact,' *American Psychologist*, XV (1960), pp. 128–34, and 'Age and productivity among scientists,' *Science*, CXXIII (1956).

17. These interviews covered some twenty administrators who were contacted in the nine companies in our total sample.

18. op. cit. (1956), p. 26.

19. 54 per cent twenty-one years plus, compared with 33 per cent sixteen to twenty years.

20. op. cit. (1966), p. 196.

21. D. C. Pelz, 'Motivation of the engineering and research specialist,' *Improving Managerial Performance*, A.M.A. General Management Ser. No. 186 (1957), pp. 25–46.

    We cannot be sure, of course, that greater interest increases productivity. It may be that success leads to a heightening of interest, and failure to a decline.

22. Moreover, their productivity peaks somewhat later and is greatest in their forties. 'Nobel Laureates in Science: Patterns of Productivity, Collaboration, and Authorship,' *A.S.R.* (1967), pp. 391–403.

23. op. cit. (1966), p. 202.

24. ibid., p. 209.

25. For a related discussion, see B. G. Glaser, op. cit. (1964).

26. Alternatively, we might employ the notion of 'role-bargain' discussed in Chapter 6.

27. D. C. Pelz and F. M. Andrews, op. cit. (1966), Chapter 2.

28. ibid., p. 10.

29. ibid., p. 22.

30. ibid., p. 22.

31. ibid., p. 24.

Chapter 8

Conclusions and Discussion

SCIENCE AND INDUSTRY

With the increasing application of science in advanced industrial societies, the interactions between science and industry are of growing significance, not only for industry, but also for science. In the past, science has been characterised by substantial autonomy, a relatively high degree of freedom from political and economic constraints. Its more recent closer interactions with industry have highlighted the distinctive characteristics of these quite different spheres of activity. Indeed, it has been a marriage of convenience. Industry was slow to embrace science, and many scientists have viewed industry with lofty disdain. The passion has come later—the exacerbation of industry on occasion with the 'prima donna' behaviour of some of its scientists that it cannot do without yet cannot really understand.

The conflicts and strains between industry and science are to some extent inevitable. Each is a distinct system of behaviour. They are different games, each with its own goals and rules. And you cannot play one game by the rules of another. In science, the goal is public knowledge, and those whose contributions to knowledge are accepted are rewarded by recognition, esteem, and honours. The goal of the economic system is the production of marketable goods and the rewards are above all monetary.

This is, of course, an oversimplification. Successful role performance in the economic system may carry a variety of rewards beyond income and security, including power and influence, prestige and recognition. But it is prestige evaluated by economic criteria, and for performance which contributes to the economic survival and growth of the organisation. The industrial scientist who is highly rewarded for his contribution

163

to company success may enjoy a scientific reputation. But this is a 'spin-off' from his main activity. And he must first and foremost play the rules of the economic game if he is to achieve its rewards. Similarly, the successful academic scientist may achieve a measure of wealth and influence, but again this depends upon his playing the academic game according to its rules. Of course, he may switch roles, or play both games at once, through consultancy or directorships, on the basis of his scientific eminence. But this is not easy. Success in the one game may contaminate recognition in the other. The amateur in sport must avoid any suspicion of monetary reward if his amateur status is to remain intact.

The successful performance of any role requires not only skills and capacities, but also commitment. The activities must be experienced as rewarding, and must meet the needs of the actor if he is to be adequately motivated. The academic scientist acquires more than knowledge, skill and capacities. Through exposure to the norms and values of science, he has accepted them as his own, and has modelled himself on the 'idols' of science. And the more he has identified with science and scientists, the more he will be motivated to seek the rewards of science through recognition, since only thus will his identity as a scientist be sustained. And it is when he finds himself in a role with which he does not identify and which does not offer the rewards he seeks, that he experiences strains and dissatisfactions.

But, as we have seen, not all with degrees in science are scientists in this sense—'public scientists'. Moreover, the majority of public scientists seek an academic role, more congruent with their identities, in which they may more successfully achieve the goals which they have come to value and the rewards which follow from their successful achievement. In fact, the universities which as one of their functions produce the next generation of academic scientists, also graduate many who are *not* dedicated to the advancement of knowledge. And it is these in particular who seek employment in the expanding opportunities in industry.

THE UNIVERSITIES AND INDUSTRY

The big expansion in the demand for science, and the related

164

expansion of the universities has been a function of the increasing applications of science—in industry, and in defence. Yet the institutions for the education of scientists have changed little since the days when science was primarily an autonomous academic activity. The future industrial scientist is still taught in universities where the role models held up for emulation are successful academics, and where the success of the student is judged in academic terms. It is not surprising that those who are most successful in the academic game should become proselytised and tend to identify themselves with the very activities for which they have been rewarded. What is lacking in the traditional university is the existence of alternative idols for emulation—men who have been notably successful in the very activities in which an increasing proportion of science graduates will be employed.

The analogy here is with the professional schools, in medicine and architecture. Particularly in medicine, the student is exposed to the influence of eminent professionals. But even here, there are problems since it is the consultants and specialists who are the most prestigeous. Hence the trend towards the development of colleges of general practice. The nearest to such professional schools are the recently created technological universities, where the emphasis is on the industrial applications of science, and the predominant pattern is sandwich courses, in which periods of academic study alternate with practical experience.

But such solutions are not without difficulties. In medicine, a high-prestige occupation, the successful consultant is probably higher in status than the successful academic. But a comparable situation is unlikely even in a technological university. The roots of this problem are deep. Western industrial societies are still permeated with essentially aristocratic values, where industry and trade are seen as less estimable, and where pure scholarship is still very highly valued. The situation is changing, but the relatively lower prestige of practical affairs is reflected in the lower prestige of engineering compared with science, and of pure science compared with applied.

The professional schools, whatever the specialism, face other pervasive problems. Professions apply knowledge. Professionals need to be not only knowledgeable but also skilled in application. And the motivation and commitment to advance knowledge is not the same as that required for its application. They are different games with different rules and rewards, played to different audiences. The dilemma is how to promote both games at once in the same institution? Advances in medicine have sprung from advances in biochemistry, cytology and other fundamental disciplines. Will it be possible for the technological universities to attract scholars who will establish the academic reputation of their institution, so that they become centres of academic excellence, without which their position in the university hierarchy will remain inferior? In so doing, will the growing concentration of such staff and students subvert their original and distinctive aims? Or will they be able to achieve a distinct and independent prestige, for their success in producing alumni who will make their mark in practical affairs and thus bring renown upon the institutions who trained them? Or will it be possible to evolve knew kinds of hybrid institution in which both kinds of excellence receive recognition and reward? How far, for example, in the technological universities, is the professional role clearly perceived and esteemed? For until it is, there is the continuing danger that industrial scientists will be failed academics.

However, we would wish to reiterate that the influence of the universities has been exaggerated. As we argued in Chapter 3, the simple hypodermic model of the university injecting academic values into passive students is over-simple and does not correspond with the facts. It is a minority of students who become academics. And the evidence suggests that they do so at least in part *because* they are attracted to the academic life.

SCIENTISTS IN INDUSTRY

The problems of relating science to industry do not depend on changes to the educational system alone. The emergence of high prestige institutions in which professional excellence is stressed would certainly help, though such institutions would be faced with the difficult problem of ensuring that there was excellence in both the pursuit of fundamental knowledge *and* in its application. Even with a shift in the values of society towards greater prestige for professionals, there is still the problem of the unfavourable image of industrial employment, which however exaggerated and distorted it may be, appears to have some substance in fact. And it is this, we have argued, which is a major factor leading the more able student to prefer a career in a university, *and to 'become' an academic.*

If our researches have failed to find among many science graduates the kinds of strains and conflicts which could be predicted from the lack of congruence between the values and norms of industry and science, there is ample evidence to suggest that industry needs to take a long hard look at the way in which it is using scientists. The management of professionals presents special problems. The literature on industrial sociology is increasingly stressing the need for participative management and to provide opportunities for self-actualisation. Such managerial strategies are even more important for professionals. The process of differential socialisation and selection ensures that by and large the more committed scientists are not recruited to industry. But is also results in the failure of industry to attract the more able student, though we must beware of attaching too much importance to degree classification as a predictor of subsequent success. If industry is to be presented as providing opportunities which are as challenging, exciting, and rewarding as those of fundamental research, it will need to heed the evidence on the under-utilisation of skills, and the absence of opportunity for intellectual stimulus. As we said in Chapter 6, there is the danger that the universities may become the whipping boys for administrative weaknesses in industry. We cannot, of course, be sure that our firms are representative of British industrial research. But the substantial differences between firms in our sample are evidence enough that all is not well with some.

We are not suggesting that all the strains and tensions between science and industry should be ironed out. Strains and conflicts may be challenging and productive. They may also be ennervating. We have tried to explore in a preliminary way some of the strategies by which scientists may be more productively accommodated to roles which are potentially stressful.

THE SOCIOLOGY OF SCIENCE

We embarked on this study primarily as an exercise in the

sociology of science. This is an embryonic area of study with no clearly demarcated boundaries. Nor have we, in this research, or in our broader review of the literature, explored all the topics which could be subsumed under this heading. Rather we have concentrated on applying the perspectives of the sociology of occupations and the sociology of organisations to an exploration of the increasingly close articulation between science and industry. In doing so, we have looked systematically at occupational socialisation and selection, and have applied role analysis to exploring the meaning of work, its motivations, strains and satisfactions. We have found evidence of role strain resulting from a lack of congruence between roles and identities, both from the conflicts between the needs of some scientists and the activities and rewards of industry, and from the frustration of opportunities for self-actualisation through the limited opportunities for the exercise of skills and autonomy.

But it is not easy to relate this to the growing literature on the sociology of science. This has been largely confined to the study of the scientific community, to the exclusion of the growing volume of scientific activity located in industrial research laboratories. To exclude industrial science would be to neglect a sizeable part of the total science activity of advanced industrial societies. Yet there is sense in making a distinction. Industry and science *are* two distinct social systems, pursuing different goals. Each game can only be played successfully if the players know which goal they are aiming for and what the rules are. It is when we move from the level of systems analysis to role analysis that we become particularly aware of the discontinuities between science and industry. Some of the confusion is semantic. We use the same word 'scientist' to describe the players who have identified themselves with the academic game of promoting knowledge which is the goal of the social system of science, and those who are willing and even prefer to play a different game, whether as professionals concerned rather with the application of knowledge, or with an organisational career in management. All are labelled scientists, yet they are very different.

The sociology of science could, from this perspective, choose to ignore those who are not public scientists, except in-so-far as they are products of the same socialisation process that produces public scientists. It must certainly, in future, differentiate between the various scientific roles and identities. It can hardly ignore the existence of public scientists engaged in the more basic research in industry. And it is here that the discontinuities between science and industry are most marked.

One particular issue of significance for the sociology of science (as well as for society) is the implications of industrial research for scientific communication—a process which is of key significance for the social system of science. We have already referred to the hindrances to publication experienced by scientists in industry. Many argued that any such restrictions do not result in any loss to science:

'There is probably no serious failure so far as reliable knowledge is concerned. Industrial research organisations encourage the publication of fundamental or basic work. It is usually possible to obtain patent protection for any commercial application of such work. In the applied field, it is sometimes difficult to obtain a valid patent, and information of use to competitors would then be withheld, but this is likely to be of technological rather than scientific interest.'

But another thought that secrecy went beyond protecting know-how and resulted in wasteful duplication:

'There is often needless secrecy which is frequently as damaging to the firm holding the secret as to a competitor and results in duplication of work within that firm, not to mention duplication between one firm and another, which viewed on a worldwide scale is wasting human effort. On the other hand many, possibly most, of the most closely guarded secrets are matters of technology and scientifically speaking are not worth knowing.'

It was also argued that a more widespread sharing of information would be economically profitable:

'Industry over a wide range is concerned with making a profit as well as a product. To make a product economically and to invent new products requires scientists to understand the process in general terms. Sharing such information will inevitably increase the profitability of firms who obtain knowledge freely, without paying for research. However, if such knowledge were spread freely the overall profitability of the whole industry must increase as there is little need for the same piece of research to be carried out twice in secrecy.'

Finally comes a most interesting argument, that there exists a network of information exchange, analogous to the 'invisible colleges' referred to earlier, which by-passes the normal channels of publication.'

'Yes, to some extent but there is a vast exchange of research information in industry which by-passes publication channels, and which is reasonably international. However, it is necessary for a company's R/D dept. to acquire prestige in successful processes (more than pure sciences) to be accepted in the "club". The scientific journals are not very interested in the results of most industrial research, except that of an academic nature. There is a lot of information in patents, once you become practised at interpreting them. Publications do not prevent duplicating even in the academic world. Journals are full of trivia as it is. There is some excuse for this in the academic world where success is measured by publications. Industrial research is frequently very detailed in its investigation of those trivia of science which are adaptable to industry. How far is the publication of this work justified? There is a better case for making it available to mechanical searching methods.'

The extent to which the industrial milieu hinders publication is then somewhat problematic. American evidence indicates quite dramatic differences between the contribution of industrial and university scientists to the growth of science, as measured by the criterion of published papers. Such evidence shows that in 1948, 54 per cent of chemists with a PhD were located in industry, but they published only 19 per cent of the papers, compared with the 64 per cent of the papers contributed by the 33 per cent of PhD chemists in universities.[1] More recently, Price has quoted figures of 70 per cent of all scientists in industry producing only 2 per cent of scientific papers.[2] Secrecy may certainly result in wasteful duplication and in the failure to utilise fully by other companies those findings which a firm does not wish to exploit. But it must also be remembered that applied

170

research is in the long run parasitic on the growth of basic scientific knowledge. And in this sense, industrial secrecy and hindrances to publication may hamper the growth of science on which industrial application in the long run depends.

FURTHER RESEARCHES

Inevitably, in any exploratory project of this kind, it is easy to look back and to see gaps and inadequacies. Some areas were deliberately omitted. We did not, for example, study the organisation of research teams, and the implications of disciplinary compared with project teams. We suspect too, that our typology of scientists is too crude, and that we have not picked up all the subtleties and variations in orientations to work, its meanings and satisfactions. It would be useful, for example, to explore more fully than we have done, the existence of a distinctive professional identity among industrial scientists, with a stress on the importance and rewards of application—that is, a 'technological' orientation. Indeed, apart from Prandy's work, little has been done on scientists in this country to explore the professionalisation of the occupation.

The measures which we used for the productivity of scientists too, were admittedly inadequate. More use could be made of scientists' judgements on the effectiveness of their work. Moreover, successful projects may be incorporated in reports internally circulated. Furthermore, we made no attempt to measure the quality of output. Some study of its potential value for production, and of innovative contributions, is clearly required.

Although we have been able to throw some light on the factors influencing the development of various scientific identities and the motivations of scientists, much remains to be done. Work is currently under way under this direction by one of the authors at Bath University, into the factors in the university experience which influence the development of distinctive attitudes and values among scientists and engineers. It is hoped that this may be able to isolate the specific influence of sandwich courses as well as contributing to a fuller understanding of the process of occupational socialisation and selection.

Finally, a caution. It is not the task of the social sciences to provide simple recipes for action. It is to be hoped that the exploratory studies reported in this book may both stimulate discussion within universities and industry on their implications, as well as encourage further research. But it must be remembered that there are no certainties in science. The currently received paradigms may be overthrown within a decade. Application requires painstaking development work. And sometimes a process works although its scientific basis may be flimsy, or a theoretical solution may not work in practice.

NOTES

1. W. O. Hagstrom, op. cit. (1965), p. 38, Table 3.
2. See M. Goldsmith and A. MacKay (eds.), *The Science of Science* (1966), p. 258.

APPENDIX 1: The Industrial Survey

Obtaining the Sample

An initial letter and preliminary questionnaire were circulated to twenty firms. These were selected from *Industrial Research in Britain* (London: Harrap Research Publication, 1962). The larger firms were chosen where it was thought that chemistry was a major research area. The preliminary letter explained our research project and made two requests. First, that a brief questionnaire be completed to provide information about the number of employees, scientists, technologists, type of research, and money allocated to various activities. Second, a request to interview five or six research chemists or administrators was made. These preliminary interviews were to be short, around half an hour. Of the twenty firms, only nine agreed to the interviews, and one agreed, after much persuasion to allow questionnaires to be distributed, although no interviews.

This method of finding a research sample is, of course, not satisfactory. Ideally, we would have liked to choose our sample from the total universe of industrial chemical research laboratories in this country, controlling for size and type of research engaged on. Unfortunately this kind of data is not available, and obtaining it would constitute a research project of its own.

We did consider approaching one of the professional associations, and although this would have given us a sample of chemists, we were particularly concerned to gain information only from chemists in a restricted number of companies as we wanted to limit the number of variables. It was also felt that scientists who joined associations or learned societies might be significantly different to those who did not join. We proceeded with the interviews in seven research laboratories, and at the same time continued correspondence with two very large companies in the hope that they would be persuaded into co-operation. Altogether we carried out sixty interviews, and ideas derived from these were incorporated into a pilot questionnaire. This was then sent to two smaller firms in the sample for a test run.

Because we then felt we were on the margin of an adequate sample, we went back to the Harrap Directory, and chose twelve additional firms. The letter contained a detailed outline of our project, an offprint of an article we had published in the *New Scientist*, and a request for a sample of chemists to whom we could distribute our questionnaire, copies of which were included. Initial replies from two firms were very encouraging. In fact, in both, permission to proceed was given from the research directors. Unfortunately, these decisions were later reversed, presumably by some higher company authority.

Our next step was to turn to Stubbs Directory. From this, we abstracted nearly fifty names of companies which were listed under manufacturers of chemical products. All of these firms were in the London area. Only three additional firms were willing to co-operate. We now had, therefore twelve firms willing to distribute questionnaires. However, three companies, in two of which interviews had been carried out, wrote to say they had changed their minds about the research. A further company said they were willing to co-operate, but only on a limited basis. They wanted us to cut back the sample of their chemists to forty or fifty.

In order to increase the size of our sample, we decided to consult the 'Directory of British Scientists'. From this, we obtained the names and addresses of over 200 chemists working for two major companies. We also selected 150 more names and addresses from a variety of companies. This was an extremely wasteful method of obtaining a sample, as the Directory often did not specify whether the chemist was engaged on research/development. In fact, as many as half of those chosen from a wide range of firms completed an additional questionnaire indicating that they were not engaged in actual work that was of interest to us. The result was that, although the response rate from firms which had agreed to co-operate in the research was mainly over 70 per cent (after two reminder letters to non-respondents), the response rate from the second method was 50 per cent.

The result of industry's generally poor response, and the equivocation of those who initially agreed, is that the final return of respondents was 422 (68 per cent of all potential respondents, and 56 per cent of all questionnaires distributed). This figure was further reduced to 403 since nineteen respondents had completed the questionnaire in such a way as to render it unusable; the final usable percentage was therefore 68 per cent.

Subsequent to the questionnaire survey, a further series of thirty-four taped interviews lasting approximately one hour was carried out.

Barriers to Research

We have discussed our problems in some detail to emphasise the difficulties of this kind of research. The reluctance of companies to co-operate is understandable. They are naturally anxious to avoid disruption to the work of valuable personnel, although it was expected that the questionnaire would, in general, be completed outside the firm's time. Some firms claim to have received a number of similar requests in recent months. It is clearly difficult for the firm to judge the merits of each enquiry, although the subsequent absence of publications on research of this kind indicates that many such inquiries failed to reach a successful conclusion. We suspect too, that the fact that the project originated from a polytechnic, and not from a university did not help.

A more fundamental reason may have been the belief in industry either that social scientists have little to offer, or that the problems were not researchable, or that they did not exist. There was certainly some evidence for this last view. There was also a possible fear that the investigation might itself generate disturbances in the laboratories by increasing the awareness of problems. But this was 1964–5. And the climate of opinion in industry has changed since then. Some of the largest companies are increasingly aware of the potential contribution of the social and behavioural scientists. Some are employing social scientists to monitor research findings, and to explore their implications for industry. Moreover, since some of the findings have become known, the authors have received a growing number of requests to talk to firms and to societies about their researches.

It goes without saying that we understand the reluctance of firms to assist in such inquiries, and are most grateful for the generous help given by the firms who co-operated and the many scientists who completed questionnaires with such care. We realise too, that social scientists must work patiently for greater understanding and acceptance. We hope that the contribution offered in this book, despite its shortcomings, will go some way to increase this acceptance.

APPENDIX 2: The Industrial Questionnaires

Survey of Scientists in Industry

1. What is the title of the position you hold at present in the company? (e.g. group leader, section leader). Could you also give name of department/section.

   ....

2. How many graduates are directly subordinate to you in your present position? ....

3. For how many years have you been in industrial research/development? ....

4. For how many years have you been working for your present company? ....

5. The technical work of scientists covers a broad range of activities. What percentage of your time is directed toward each of the following purposes? (Please enter nearest 5 per cent.)

   1. Research (discovery of new knowledge either basic or applied):
      a. General knowledge relevant to a broad class of problems. ....%
      b. Specific knowledge for solution of particular problem(s). ....%
   2. Development: (improvement of existing products and processes; translating knowledge into useful form). ....%
   3. Technical services to help other people or groups; consultation, 'trouble-shooting', testing. ....%
   4. Other (please specify): ....%

   Total should equal 100%

6. From the following groups of people, please mark three in whose eyes it is important (to you) for you to appear well—for them to have a good opinion of your performance and accomplishments. (Please tick three.)

   1. Top executives in the parent organisation. ....
   2. Research administrators in the R/D division. ....

176

3. Professional colleagues within the R/D division. ....
4. Professional colleagues elsewhere in my field. ....
5. Respected friends outside of my field. ....
6. My family. ....

7. In your present research position, are you usually allowed—(Please tick only one that comes nearest to describing your present amount of autonomy.)

   1. to *select* the project on which you will work? ....
   2. to *assist* in the selection of projects on which you will work? ....
   3. to *decide* how you will tackle the projects which have been assigned to you? ....
   4. to *assist in the decision* on how best the project you are working on should be tackled? ....

8. Which of the above four categories (in Q. 7) would you like to have ticked? ....

9. If the answer to Q. 8 is different to that of Q. 7, how dissatisfied are you with the present arrangements? (Please tick one of the following.)

   1. .... slightly dissatisfied;   2. .... dissatisfied;
   3. .... highly dissatisfied.

10. We would like to know something about the number of scientific papers you have had published, the number of patents stemming from your work, the number of addresses you have made at scientific meetings, and the number of books you have published. In each case, we would like to know the total number, and the yearly average over the last five years.

	Total	Yearly average in last five years
1. Papers published.	....	....
2. Papers written.	....	....
3. Patents derived from work.	....	....
4. Addresses to conferences, etc.	....	....
5. Books published.	....	....

11. There are a number of factors which may hinder your publication rate. Do you think that your contribution to public science through publications is hindered in any way?

   1. Yes ....; 2. No ....

If yes, which of the following statements comes nearest to the reasons why you do not publish more papers? (Please tick all those which you feel contribute to your not publishing as much as you could.)

1. The organisation for which I work does not allow every research finding to be published. ....

2. I write more than I publish; I am too critical and send very little off for publication. ....

3. After a project is finished to the satisfaction of the company more work would be necessary to make it adequate for publication; I am not allowed that during normal working hours. ....

4. Most of my important findings are patented, and therefore the need to publish is diminished. ....

5. My family life occupies the time I would need if I were to increase my publication rate. ....

6. My other (leisure) pursuits do not leave me enough spare time to increase my publication rate. ....

7. The company insists on complete coverage by patents before publication; by this time, I am no longer interested. ....

8. My skills are underemployed on work that is not publishable. If I had different publishable work, my rate would increase. ....

9. Other: (please specify) ....
....

12. About what percentage of projects on which you are engaged or involved in end up successfully, i.e. *a scientific solution to the problem is found, although not necessarily developed as commercial products?*
% 100 ( ); 99–80 ( ); 79–60 ( ); 59–40 ( ); 39–20 ( ); 19–0 ( )

13. There are bound to be a number of unsuccessful attempts to complete a project as far as finding a scientific solution is concerned. How would you describe *your particular success rate* for the kind of projects on which you are engaged?

        1. Higher than is normal. ....
        2. Slightly higher than is normal. ....
        3. About the normal. ....
        4. Slightly lower than is normal. ....
        5. Lower than is normal. ....

14. Regardless of how you have answered Q. 13, you may still think your research performance could be better, but a number of factors are hindering you. Do you think that your contribution to the scientific solution of problems is hindered in any way?

1. Yes; my effectiveness is considerably ($> 50\%$) less than it could be. ....
2. Yes; my effectiveness is somewhat ($< 50\%$) less than it could be. ....
3. No; I am being as productive as I could possibly be. ....

15. In your present job, when a research project is terminated, *who* takes the decision to have it terminated? If this is different on different occasions, who *mainly* takes the decision, e.g. company management, research management, immediate supervisor, myself?

....

16. Do you have the impression that the area of research on which you are mainly engaged is: (Please tick one)

   1. expanding quickly? ....
   2. expanding slowly? ....
   3. not changing? ....
   4. contracting slowly? ....
   5. contracting quickly? ....

17. Roughly how many scientific meetings have you attended in the last year? 0 ( ); 1–3 ( ); 4–6 ( ); 7–9 ( ); 10–12 ( ); 13–15 ( ); More than 15 ( ).

18. Some scientists would like to attend more scientific meetings, others would not. Their reasons for doing either can be different. Which of the following statements applies to you?

   1. I do not want to attend any more meetings than I attend now. ....
   2. I would like to attend more meetings, but I do not because:

      a. of the time and distance. ....
      b. the organisation for which I work does not encourage R and D staff to attend many meetings. ....
      c. other (specify)..

      ....

19. Roughly how much of your work-week is 'free-time', i.e. when you are more or less allowed to pursue ideas that have not been specifically sanctioned? (Please give percentage answer.) ....%

20. Would you like to have more of this 'free-time' allowed to you?

   1. Yes ....
   2. No ....

21. Smith is an industrial research scientist. Below are a number of situations in which Smith behaves in a certain way. We would like you to think about Smith's behaviour, and say whether *you* agree/ disagree with it, *in the sense that you would do the same or not in his situation*. For your answer, please use the following code: 1 = total agreement; 2 = moderate agreement; 3 = uncertain, mixed feelings; 4 = moderate disagreement; 5 = total disagreement.

Research scientist Smith is offered another industrial position where he has more freedom in his job, and the research is very similar. The job is however a fair distance away, say over 200 miles, although still in the UK.

1. Smith, who is married, accepts the offer. ....
2. Smith's parents will be slightly worse off financially; Smith accepts the offer. ....

Research scientist Smith has just made a contribution to science which he would like to publish. His company, however, urges him not to do so, since, his findings may be of use to competitors.

3. Smith is offered a financial bonus for his work; he accepts the bonus and decides not to push for publication. ....
4. Smith, a married man with two children, goes ahead and publishes, although he knows he runs the risk of dismissal. ....

Research scientist Smith, *who is now in his early thirties*, has been working for a company for over five years. He is offered an administrative position with more status and a higher salary, but with little scope for further research.

5. Smith, who is single, accepts the offer. ....
6. Smith, who is married with no children, accepts the offer. ....

22. People obtain satisfaction at work from different sources. Please look at the following three statements, and say how far you agree with each. Code: 1 = complete agreement; 2 = agree; 3 = uncertain; 4 = disagree; 5 = complete disagreement.

1. One of my main satisfactions comes from the interest and sometimes the excitement of solving scientific problems. ....
2. One of my main satisfactions is from seeing the results of my research efforts incorporated into a company product which sells well. ....
3. One of my main satisfactions comes from publishing a paper which is well received in the scientific world. ....

You may have scored all three the same, or two the same. We would like you to circle the number of the item (1, 2, or 3) which in fact gives you the most pleasure.

Please read carefully

23. Listed below are a number of conditions and auxiliary services that may contribute to the effective performance of the work in your laboratory. We would like you to look at these conditions and answer three questions related to each condition. In column A please rate each item on how satisfied you are with the adequacy of these conditions in your R/D organisation at the present time. In column B please say how you think most research administrators in your R/D organisation would answer the first questions, i.e. how satisfied they are. In column B *we are not asking you to state a fact about what is the actual opinion of most research administrators, but merely how you think they would answer.* In columns A and B please use the following code: 1 = highly satisfied; 2 = satisfied; 3 = neutral feelings; 4 = dissatisfied; 5 = highly dissatisfied; 6 = does not apply. In column C please rate each condition on how important a contribution you think it makes towards the productivity of R/D scientists. In this column please use the following code: 1 = very important; 2 = important; 3 = mixed feelings about importance; 4 = unimportant; 5 = very unimportant. Please leave column D blank.

	A Your satis-faction	B Research admin's satisfac-tion	C Impor-tance to research	D
1. The salaries of R/D chemists.	....	....	....	....
2. The amount of free-time allowed for private research.	....	....	....	....
3. Quantity and quality of assisting and supporting personnel (technicians, etc.).	....	....	....	....
4. The amount and security of research and development funds.	....	....	....	....
5. Prospects for promotion up a scientific career ladder.	....	....	....	....

6. The amount of in-
   fluence research
   workers have in
   choosing their re-
   search projects. ....

7. The general condi-
   tions of building—
   heating, ventilation,
   etc. ....

8. Research workers
   influence over their
   hours of work. ....

9. The way in which
   projects are ter-
   minated. ....

10. Attention given by
    management to re-
    search personnel
    suggestions and
    complaints. ....

11. Publications policy. ....

12. Patents policy. ....

13. Prestige of labora-
    tory in the scientific
    world. . . . .

14. The way in which
    highly productive
    research workers
    are rewarded. ....

15. The inter-personal
    relations between
    research staff and
    company manage-
    ment. ....

16. Long-term plan-
    ning of research
    programme. ....

17. The opportunities
    to attend scientific
    meetings and con-
    ferences. ....

18. Recruitment policy
    into research staff. ....

19. The over-all organ-
    isation of research
    (team structure;
    departmental
    division). ....

20. The degree of su-
    pervision over your
    main work.   ....
21. The extent to which
    your capacities and
    skills are employed.   ....

24. Are you a member/associate, etc., of any learned society, professional
    association, or professional trade? (If so, please state the initials,
    e.g. ARIC.)

25. If there were a General Election tomorrow, and you could vote for
    any of the following parties, for which would you vote? For which
    did you vote at the last election? How strongly do you believe in the
    principles for which this party (the one for which you would vote now)
    stands? (Use code: $1 =$ very strongly; $2 =$ strongly; $3 =$ not
    strongly.)

	Would vote for	Did vote for	Believe in
	....	....	....
1. Communist	....	....	....
2. Conservative	....	....	....
3. Labour	....	....	....
4. Liberal	....	....	....
5. Other	....	....	....

26. Please state how far you agree/disagree with the following state-
    ments—

	agree 1	2	3	4	disagree 5
1. There is nothing unpatriotic about working in a foreign country, since a scientist's loyalty is towards science.	....	....	....	....	....
2. The values of science some-times conflict with the jobs governments require scientists to work on.	....	....	....	....	....
3. The policies of national govern-ments frequently interfere with the exchange of scientific knowledge across national boundaries.	....	....	....	....	....
4. I think of myself as a scientist first and a member of a nation second.	....	....	....	....	....

27. In some companies, there is a dual career ladder, i.e. a person can stay in research until his early thirties, and then switch to research administration, or administration in another company section, or he can stay directly involved in research. Positions on both these career ladders are more or less equal in terms of salary, status, and promotion chances.

    1. Does such a dual career ladder exist in your company?

           Yes.   ....

           No.   ....

    2. If yes, do you intend to take the first option, i.e. research administration?

           Yes.   ....

           No.   ....

    3. If yes, at roughly what age would you like to move into research administration?   ....

    4. If no, i.e. there is no dual career line, we would like to know your career intention and prospects. Please tick one of the following which comes nearest to representing your situation. (If none of these fairly represent your situation, please complete 5.)

        1. If I can move into research management, I shall do so; in fact, I think my chances of being made such an offer are very good.   ....

        2. If I can move into research management, I shall do so; but my chances of being made such an offer are not good.   ....

        3. I intend staying in research and not moving into research management if I can avoid it.   ....

        4. I shall attempt to obtain another research post in:
           a. another industrial firm.   ....
           b. a university, C.A.T., etc.   ....
           c. a government laboratory.   ....
           d. other; ....

        5. I intend moving out of R/D completely and into:
        ....

28. If the company for which you work does not have a dual career line, do you think that it ought to have one? 1 Yes.........; 2 No........
What is the main reason for your answer?.......................................
....

Personal Details Form: We would like you to give us some details about yourself and your parents. These are not to enable us to identify you, but are merely to be used in the inquiry as control variables.

A. Sex: Male:.......... Female: ....

B. Age: 20–24...........; 25–29...........; 30–34...........;
35–39...........; 40–44...........; 45–49...........;
50–54...........; 55–59...........; 60+ ....

C. Present 1000 or less ....; 1001–1249...........; 1250–1499...........;
Salary 1500–1749...........; 1750–1999...........; 2000–2249...........;
2250–2499...........; 2500–2749...........; 2750–2999...........;
3000–3249...........; 3250–3499...........; 3500+ ....

D. Previous types of school attended: (e.g. elementary, senior, grammar, secondary modern, technical, public, university—please give all that you attended).
....

E. Qualifications: (please give details including grades, and the year in which they were obtained—do *not* give G.C.E., or equivalents).
....
....

F. Father's occupation: (now, or when last employed—please give as full details as possible).
....

G. Father' education:
All types of Schools attended:..
Qualifications: ....

H. Name of country in which you resided when you were young: up to 15 years of age. If more than one, give the one in which you resided most.
....

I. Single...........: Married with 0...........; 1...........; 2...........; 3...........; 4+...........: children.

J. What is the sex of your children: eldest first, and so on.
....

K. To which system of faith/morality did you subscribe as a child? To which system do you and your parents (did if deceased) subscribe *now*? (Please use code: 1 = utmost conviction; 2 = high conviction; 3 = some conviction; 4 = nominal.)

185

	You as child	You now	Father now	Mother now
1. C. of E.	....	....	....	....
2. Other protestant.	....	....	....	....
3. Roman Catholic.	....	....	....	....
4. Judaism.	....	....	....	....
5. Agnosticism/ Atheism/ Humanism.	....	....	....	....
6. Other:	....	....	....	....

L. Please give details of all posts held before *your present position.*

Major position	Type of location, e.g. university, industry, government, etc., and name of organisation.	Position held	Main duties, e.g. basic, applied research, development, administration	Approx. no. of years in this position
1st				
2nd				
3rd				
4th				
5th				
6th				

Below are questions which require written answers. We realise we have already taken up some of your time, but we would appreciate some brief which replies to these questions.

Do you think that there is a failure in industry to publish knowledge which should be available to other scientists?

It is often suggested that there are conflicts between the values of science and the values of industry, and that scientists working in industry sometimes experience these conflicts in the form of difficulties and strains. Do you think some scientists in your R/D organisation experience such difficulties? If so, could you give us a brief example of this.

Any other comments that you think might be of importance, or are related to the questionnaire. Please continue on the back if necessary.

Thank you very much for your co-operation.

Survey of Research Administrators

1. What is the title of the position you hold at present in the company? (Please give full details, e.g. Head, Deputy-Head, R/D department or section within this.)
   ....

2. For how many years (to the nearest whole number) have you been in your present position?—present company?—industrial research/development?

Present position	*Present company*	*Industrial R/D*
....	....	....

3. The work of a research administrator covers a wide range of activities. How much of your time do you spend on the following activities during a typical week? (Please answer to nearest 5 per cent.)

   1. Administration (paper work) in office.   ....
   2. Bench work.   ....
   3. Direct discussion in the laboratories with researchers on their project problems.   ....

187

4. Discussion in office with researchers on their project problems. ....
5. Discussion with other research administrators. ....
6. Discussion with members of top management. ....
7. Discussions with members of other departments. ....
8. Other: (please specify):.. ....

4. From the following groups of people, please mark three in whose eyes it is important (to you) for you to appear well—for them to have a good opinion of your performance and accomplishments. (Please tick three.)

1. Top executives in the parent organisation. ....
2. Research administrators in the R/D division. ....
3. Professional colleagues within the R/D division. ....
4. Professional colleagues (scientists) elsewhere in my field. ....
5. Respected friends outside of my field. ....
6. My family. ....

5. Smith is an industrial research scientist. Below are a number of situations in which Smith behaves in a certain way. We would like you to think about Smith's behaviour, and say whether you agree/disagree with it, *in the sense that you would do the same or not in his situation.* For your answer, please use the following code: 1 = total agreement; 2 = moderate agreement; 3 = uncertain, mixed feelings; 4 = moderate disagreement; 5 = total disagreement.

Research scientist Smith is offered another industrial position where he has more freedom in his job, and the research is very similar. The job is, however, a fair distance away, say over 200 miles although still in the UK.

1. Smith, who is married, accepts the offer. ....
2. Smith's parents will be slightly worse off financially; Smith accepts the offer. ....

Research scientist Smith has just made a contribution to science which he would like to publish. His company, however, urges him not to do so, since his findings may be of use to competitors.

3. Smith is offered a financial bonus for his work; he accepts the bonus and decides not to push for publication. ....
4. Smith, a married man with two children goes ahead, and publishes, although he knows he runs the risk of dismissal. ....

Research scientist Smith, who is now in his early thirties, has been working for a company for over five years. He is offered an administrative position with more status and a higher salary, but with little scope for further research.

5. Smith, who is single, accepts the offer. ....
6. Smith, who is married with no children, accepts the offer. ....

Please read carefully.

6. Listed below are a number of conditions and auxiliary services that may contribute to the effective performance of the work in your laboratory. In column A, please rate each on how satisfied *you* are with the adequacy of these conditions in your R/D organisation at the present time. In column B, please say how *you think most research chemists* in your R/D organisation would answer the same question. *In column B, we are not asking you to state a fact about the actual opinions of most research chemists, but merely how you think they would answer.* In both columns A and B, please use the following codes: 1 = highly satisfied; 2 = satisfied; 3 = neutral feelings, mixed; 4 = dissatisfied; 5 = highly dissatisfied; 6 = does not apply.

	A Your satis-faction	B Most R/D chemists satis-faction
1. The salaries of R/D chemists.	....	....
2. The amount of 'free-time' allowed for private research.	....	....
3. Quantity and quality of assisting and supporting personnel, (technicians, etc.).	....	....
4. The amount and security of research and development funds.	....	....
5. Prospects for promotion up a scientific career ladder.	....	....
6. The amount of influence research workers have in choosing their research projects.	....	....
7. The general conditions of buildings— heating, ventilation, etc.	....	....
8. Research workers influence over their hours of work.	....	....
9. The way in which projects are terminated.	....	....
10. Attention given by management to research personnel suggestions and complaints.	....	....

11. Publications policy. ....
12. Patents policy. ....
13. Prestige of laboratory in the scientific world. ....
14. The way in which highly productive research workers are rewarded. ....
15. The interpersonal relations between research staff and company management. ....
16. Long-term planning of research programme. ....
17. The opportunities to attend scientific meetings and conferences. ....
18. Recruitment policy into research staff ....
19. The overall organisation of research (team structure; departmental divisions.) ....
20. The degree of supervision of researchers main work. ....
21. The extent to which researchers capacities and skills are employed. ....

7. Are you a member/associate, etc., of any learned society, professional association, or professional trade union? (If so, please give initials, e.g. ARIC.)

8. There are a number of factors which may hinder the publication rate of R/D personnel. Do you think the contribution to science through publications of the R/D personnel in your organisation is hindered in any way?

    1. Yes ....; 2. No ....

If yes, which of the following statements comes nearest to the reasons why more is not published by your company's R/D personnel? (Please tick all those which you feel contribute to the publication rate being less than it could be.)

1. This organisation does not allow all research findings to be published. ....
2. They write more than they publish; they are too critical and send very little for publication. ....
3. There is a big gap between work completed to the satisfaction of the company, and that desired by an academic journal. The pressure of work does not allow this extra work to be completed. ....
4. Most important findings in this company are patented and therefore the *need* for publication is diminished. ....

5. The scientists want to spend their free-time with their family. ....

6. The scientists are more concerned with other (leisure) activities. ....

7. By the time patents are completed, the scientists have lost any enthusiasm to write up the work for an academic journal. ....

8. More scientists could publish work, but their present work under-uses their capacities and skills. ....

9. Other: (please specify): ....
....

9. There are bound to be a number of unsuccessful attempts to complete a research project as far as finding a scientific solution is concerned. How would you describe the *particular success rate* for your R/D personnel for the kind of projects on which they are engaged?

1. Higher than is normal. ....
2. Slightly higher than is normal. ....
3. About the normal. ....
4. Slightly lower than is normal. ....
5. Lower than is normal. ....

10. Regardless of how you have answered Q. 9 you may still feel that the overall research performance could be better, and that scientific production (solution to problems) could be higher if a number of factors which hinder research were removed. Do you think the research performance could be higher than it is?

1. Yes: probably more than 50 per cent better. ....
2. Yes: but less than 50 per cent better. ....
3. No: research performance is at its highest possible. ....

11. If you have previously been directly in research work, we would like to know something about the number of papers you have had published, the number of patents stemming from your work, the number of addresses you have given to scientific meetings, and the number of books you have published. In each case, we would like to know the total number, and the yearly average in the last five years.

	Total	Yearly average in last five years
1. Papers published.	....	....
2. Papers written.	....	....
3. Patents derived from work	....	....
4. Addresses to conferences, etc.	....	....
5. Books published.	....	....

191

12. We are interested in the reasons which lead you into research management. Would you please look at the following reasons and tick the one which represents most closely the reason you had at the time of entering research administration?

    1. There was no other means of obtaining promotion. ....

    2. I mastered my previous job and needed a new stimulating position. ....

    3. I did not think scientists interests were being well represented in research management, and I wanted to improve this. ... .

    4. I thought that at that age, and experience, I would simply make a better research administrator than a research scientist. ....

    5. My research job grew as the research area grew and I found myself doing more and more administration. ....

    6. I never thought I was really suited to research; administration has always really been my line. ....

    7. Other: (please specify):...
....

13. We are interested in your career intentions and career prospects. Could you please tick one of the following which comes nearest to representing your situation. (If none of these fairly represent your situation, please complete item five.)

    1. If I can move into top management and out of research management, I shall do so; in fact, I think my chances of being made such an offer are very good. ....

    2. If I can move into top management and out of research management, I shall do so; but my chances of being made such an offer are not very good. ....

    3. I do not intend moving out of research management. ....

    4. I shall attempt to obtain another research management post in:

       a. another industrial firm. ....

       b. a government laboratory. ....

       c. other: (please specify):...
....

    5. I intend moving out of research management and into:

....

14. If you have ticked codes 1 or 2 in the previous question, Q. 13, could you now complete the following question.

A research administrator could have a number of reasons for

wanting to join top management. Briefly, could you give us your prime reason for wanting to move.

....

....

15. Thinking back to the time you joined research administration, can you remember whether you felt any strains about leaving research? Did you for instance wish you could have stayed in research? Would you have liked to have joined administration later? Were you entirely happy about the move? Did you have subsequent regrets?

....

....

....

....

....

....

....

....

Personal Details Form: We would like you to give us some details about yourself and your parents. These are not to enable us to identify you, but are merely to be used in the inquiry as control variables.

A. Sex: Male:............... Female:...............

B. Age: 20–24...........; 25–29...........; 30–34...........;
         35–39...........; 40–44...........; 45–49 ....;
         50–54...........; 55–59...........; 60+ ....

C. Present 1000 or less ....; 1001–1249...........; 1250–1499...........;
   Salary: 1500–1749...........; 1750–1999...........; 2000–2249...........;
            2250–2499...........; 2500–2749...........; 2750–2999...........;
            3000–3249...........; 3250–3499...........; 3500+ ....

D. Previous types of school attended: (e.g. elementary, senior, grammar, secondary modern, technical, public, university—please give all that you attended).

....

E. Qualifications: (please give details including grades, and the year in which they were obtained—do *not* give G.C.E., or equivalents).

....

....

F. Father's occupation: (now, or when last employed—please give as full details as possible).

....

G. Father's education:
All types of Schools attended:..... ....
Qualifications: ....

H. Name of country in which you resided when you were young: up to 15 years of age. If more than one, give the one in which you resided most.

....

I. Single .... ; Married with 0.......... ; 1.......... ; 2............... ; 3..... .....;
4+............: children.

J. What is the sex of your children: eldest first, and so on.

....

K. To which system of faith/morality did you subscribe as a child? To which system do you and your parents (did if deceased) subscribe *now*? (Please use code: 1 = utmost conviction; 2 = high conviction; 3 = some conviction; 4 = nominal.)

	You as child	You now	Father now	Mother now
1. C. of E.	....	....	....	....
2. Other protestant.	....	....	....	....
3. Roman Catholic.	....	....	....	....
4. Judaism.	....	....	....	....
5. Agnosticism/ Atheism/ Humanism.	....	....	....	....
6. Other:	....	....	....	....

L. Please give details of all posts held before *your present position*.

Major position	Type of location, e.g. university, industry, government, etc., and name of organisation	Position held	Main duties, e.g. basic, applied research, development, administration	Approx. no. of years in this position
1st	....	....	....	....
2nd	....	....	....	....
3rd	....	....	....	....
4th	....	....	....	....
5th	....	....	....	....
6th	....	....	....	....

Below are questions which require written answers. We realise we have already taken up some of your time, but we would appreciate some brief replies to these questions.

Do you think that there is a failure in industry to publish knowledge which should be available to other scientists?

It is often suggested that there are conflicts between the values of science and the values of industry, and that scientists working in industry sometimes experience these conflicts in the form of difficulties and strains. Do you think some scientists in your R/D organisation experience such difficulties? If so, could you give us a brief example of this.

Any other comments that you think might be of importance, or are related to the questionnaire. Please continue on the back if necessary.

Thank you very much for your co-operation.

APPENDIX 3: The University Questionnaire

1. What is the name of the degree for which you are at present reading?
   (Please give full details.)
   ....

2. a. What level of degree do you expect to obtain? If you think
      you might possibly obtain a higher or lower degree, please tick in
      appropriate column.
   b. If you have graduated and are pursuing a higher degree, what was
      your bachelor degree level?

	Expected degree level	Possible higher degree level	Possible lower degree level	Level already obtained
1. First.	....	....	....	....
2. Upper second.	....	....	....	....
3. Lower second.	....	....	....	....
4. Pass/Third.	....	....	....	....
5. Fail.	....	....	....	....
6. Really cannot say.	....	....	....	....

3. At what age did you specialise in the main subject(s) of your present
   degree course? At what age did you decide in principle to pursue
   your present degree course?

	Age: Years Old								
	13	14	15	16	17	18	19	20	Older
I specialised at:	....	....	....	....	....	....	....	....	....
I decided in principle at:	....	....	....	....	....	....	....	....	....

4. If you did study your degree main subject(s) at school, were you—
   (please tick one of the following.)

   1. Much more successful at it than other subjects.   ....
   2. More successful at it than other subjects.   ....
   3. About the same at it as other subjects.   ....
   4. Less successful at it than other subjects.   ....
   5. Much less successful at it than other subjects.   ....

196

5. If you did study your degree main subject(s) at school, what did you feel about the teacher(s) who taught this subject? (Please tick one of following.)

   1. I admired him much more than other teachers.   ....
   2. I admired him more than the other teachers.   ....
   3. I admired him as much as any of the teachers.   ....
   4. I admired him less than other teachers.   ....
   5. I admired him much less than other teachers.   ....

6. What types of school did you attend? Did your father attend? Did your mother attend? (*Please tick all types attended by each.*)

	You	Father	Mother
1. Public or independent school.	....	....	....
2. L.E.A. grammar school.	....	....	....
3. Comprehensive school.	....	....	....
4. Technical school.	....	....	....
5. Secondary modern/senior/ central.	....	....	....
6. Elementary.	....	....	....
7. College of further education.	....	....	....
8. U.K. university or equivalent.	....	....	....
9. Overseas university.	....	....	....

7. In many secondary schools, classes are grouped together according to students abilities, i.e. *streamed*. The most able go into 'A' stream, or its equivalent, the next able group into the 'B' stream, and so on.

   a. Was your school streamed? If so, in which stream were you?
   Yes...........: No............. A............: B.................: C...........: D...........: E.............
   b. If not streamed, were you in the top ability half of your class, or bottom half?
   Top half...........: Bottom half............

8. Which of the following reasons comes nearest to the main reason why you are studying for the above degree? (Please tick the appropriate *one*, or complete option seven.)

   1. I had always liked my degree subject(s) at school and wanted to know more about them.   ....
   2. My family encouraged me.   ....
   3. My friends encouraged me.   ....
   4. I thought this degree would help me obtain a good job.   ....
   5. I did not like the idea of work and taking a degree would enable me to put the decision off for some years.   ....

6. It was a natural continuation. ....

7. Other: (Please specify)...

....

9. In some of the serious literature on scientists, different types of scientist are frequently mentioned. Below are three such types.

Type A: dedicated to science, engaged in much research, *eager to publish.*

Type B: dedicated to science, does engage in research *not eager to publish.*

Type C: defines science as an interesting way of earning a living; preoccupied with non-scientific matters.

(i). How many of the staff that you have come into contact with in your university/college do you think are *more or less* like Type A, B, or C? How many of your friends *at* university/college are, or are likely to become *more or less* like type A, B, or C? How many friends *outside* university/college are, or are likely to become *more or less* like type A, B, or C?

Per cent

	100	90	80	70	60	50	40	30	20	10	0
University staff											
like type—A	....	....	....	....	....	....	....	....	....	....	....
like type—B	....	....	....	....	....	....	....	....	....	....	....
like type—C	....	....	....	....	....	....	....	....	....	....	....
University friends											
like type—A	....	....	....	....	....	....	....	....	....	....	....
like type—B	....	....	....	....	....	....	....	....	....	....	....
like type—C	....	....	....	....	....	....	....	....	....	....	....
Friends outside university											
like type—A	....	....	....	....	....	....	....	....	....	....	....
like type—B	....	....	....	....	....	....	....	....	....	....	....
like type—C	....	....	....	....	....	....	....	....	....	....	....

(ii). How do you feel about the members of the staff who are more or less like one of these three types? (Please tick one of the following statements for Type A, B, and C.)

	Staff members like		
I respect them:	Type A	Type B	Type C
1. much more than other staff members.	....	....	....
2. more than other staff members.	....	....	....

    3. about the same as other staff
       members. ....
    4. less than other staff members. ....
    5. much less than other staff
       members. ....

(iii). How involved with your friends at university/college are you, compared with your involvement with friends not attending university/college?

    1. much more involved with friends at university. ....
    2. more involved with friends at university. ....
    3. involvement with both groups about the same. ....
    4. less involved with friends at university. ....
    5. much less involved with friends at university. ....
    6. I have no friends outside university. ....

10. Below are possible types of job you could choose to have after you have graduated. If you obtain the degree you expect to obtain, which of these jobs do you think you will choose. (Please tick the *one* appropriate answer in *the* column that indicates the degree of certainty you have about this choice.)

	I will definitely choose this	The chances are I will choose this	I may choose this one more than any other
1. Research position in industry.	....	....	....
2. Development position in industry.	....	....	....
3. Research position in government laboratory.	....	....	....
4. Development work in government laboratory.	....	....	....
5. Stay at university/CAT to do higher degree with ultimate view to:			
a. research work in industry.	....	....	....
b. research work in government laboratory.	....	....	....
c. research/lecturing post in university/CAT.	....	....	....
6. Stay at university/CAT to do higher degree out of pure interest *or* because have not yet decided on a desired job.	....	....	....

7. Teaching in technical college/
grammar/secondary, etc.   ....
8. Other: ....

Please read carefully. If you have decided that you will work in research/ development or take a higher degree, please complete the following question. If you have decided to teach, or do non-research/development work, or not take a higher degree, omit the next question, and turn to question 12.

11. We would like you to tell us how you feel about various character- istics of research/development positions in industrial laboratories, government laboratories, and in universities/colleges of advanced technology.

In column A please score how important each of these items are to you, using the following code: 1 = utmost importance; 2 = very important; 3 = more than important, but not very important; 4 = important; 5 = not important.

In column B please state how you think these conditions are provided in typical industrial laboratories, government laboratories and in universities/CATs. Using the following code for answering: 1 = provision is excellent; 5 = provision is poor; 2, 3, 4, are intermediary grades.

	Column A	Column B		
		Adequacy of provision in		
	Impor- tance to self	Industrial labora- tory	Govern- ment labora- tory	Univer- sity/ CAT
1. Salary scales.	....	....	....	....
2. Holiday allowance.	....	....	....	....
3. Freedom to publish research.	....	....	....	....
4. Freedom to choose research subjects.	....	....	....	....
5. Technical equip- ment.	....	....	....	....
6. Social and welfare conditions.	....	....	....	....
7. Freedom to choose work colleagues.	....	....	....	....
8. A career ladder for scientists.	....	....	....	....
9. Geographical loca- tion/natural sur- roundings.	....	....	....	....

12. An organisation, A, does not allow its scientific staff to publish their findings. This organisation is located in England, and has good technical equipment, and is generally average in other factors.

Would you be prepared to work there if—after tax: (Please tick one answer to every question (i.e. 28 ticks in all.)

		Yes	Maybe	No
	The salary was always			
1.	10 per cent more than elsewhere.	....	....	....
2.	20 per cent more than elsewhere.	....	....	....
3.	30 per cent more than elsewhere.	....	....	....
4.	40 per cent more than elsewhere.	....	....	....
5.	50 per cent more than elsewhere.	....	....	....
6.	75 per cent more than elsewhere.	....	....	....
7.	100 per cent more than elsewhere.	....	....	....

An organisation, B, does not allow its research staff freedom to choose research projects, and it insists that projects be stopped when higher officials say so.

Would you be prepared to work there if—after tax:

		Yes	Maybe	No
	The salary was always			
1.	10 per cent more than elsewhere.	....	....	....
2.	20 per cent more than elsewhere.	....	....	....
3.	30 per cent more than elsewhere.	....	....	....
4.	40 per cent more than elsewhere.	....	....	....
5.	50 per cent more than elsewhere.	....	....	....
6.	75 per cent more than elsewhere.	....	....	....
7.	100 per cent more than elsewhere.	....	....	....

An organisation, C, is the *only* organisation which is prepared to employ you to do research. This organisation is located in England, and has good technical equipment and is generally average in other factors.

Would you be prepared to work there if— after tax:

		Yes	Maybe	No
	The salary was always			
1.	1 per cent *less* than elsewhere.	....	....	....
2.	2 per cent *less* than elsewhere.	....	....	....
3.	3 per cent *less* than elsewhere.	....	....	....
4.	5 per cent *less* than elsewhere.	....	....	....
5.	10 per cent *less* than elsewhere.	....	....	....
6.	20 per cent *less* than elsewhere.	....	....	....
7.	50 per cent *less* than elsewhere.	....	....	....

An organisation, D, allows and encourages its scientific staff to publish their research findings. It further allows them complete freedom over choice of project, and termination of project. It is the *only* research job you have been offered.

Would you be prepared to work there for five years if:

	Yes	Maybe	No
It was located in—			
1. Germany/France.			
2. Scandinavia.			
3. Italy/Spain.			
4. Russia.			
5. USA.			
6. an underdeveloped African country.			
7. an underdeveloped Asian country.			

13. Are any of your relatives and/or friends working in industrial or government research and/or development laboratories, or university/CATs as lecturers and/or researchers? If so, please tick the appropriate box(es).

	Industrial R/D	Government R/D	University/CAT
Friends.			
Relatives.			

14. How old are you?...............

15. What is your sex? male...............: female................

16. Have you (or did you have) any brothers or sisters? If so, please complete the following table: give information for all brothers and sisters.

Delete inappropriate sibling	Years older than me											Years younger than me										
	10	9	8	7	6	5	4	3	2	1	0	1	2	3	4	5	6	7	8	9	10
Brother/Sister																					
Brother/Sister																					
Brother/Sister																					
Brother/Sister																					
Brother/Sister																					

17. In which country did you spend the majority of your youth up to fifteen years of age?..

18. What is/or was, (if retired or deceased) your father's occupation, and qualifications, and what is (or was) your mother's occupation, and qualifications? (*Please give full details.*)

Your father's occupation ....

Your father's qualifications ....

Your mother's occupation ....

Your mother's qualifications ....

19. Could you please give a rough estimate of the gross income of your immediate family, i.e. father/mother? Answers are per annum.

1. Less than £1,000 ....;   2. £1,001–1,500 ....:

3. £1,501–2,000 ....:   4. £2,001–2,500 ....:

5. £2,501–3,000 ....:   6. £3,000 plus ....

20. It is often suggested that people belong to social classes. To which of the following do you think you will belong in five years time? Which do you think your parents belong to? (*Please tick appropriate one in each column.*)

	You now	You in 5 years	Your parents
1. Upper class.	....	....	....
2. Upper middle class.	....	....	....
3. Middle middle class.	....	....	....
4. Lower middle class.	....	....	....
5. Upper working class.	....	....	....
6. Between upper/ lower working class.	....	....	....
7. Lower working class.	....	....	....

21. To which faith/moral system do you now, and as a child, and your parents now subscribe? Please qualify your answer by using the following code: 4 = utmost conviction; 3 = strongly: 2 = moderately; 1 = nominally.

	You as child	You now	Father now	Mother now
1. Church of England.	....	....	....	....
2. Other Protestant:				
....	....	....	....	....
3. Roman Catholic.	....	....	....	....
4. Judaism.	....	....	....	....
5. Agnostic/Atheism. Humanism.	....	....	....	....
6. Other (State):				
....	....	....	....	....

22. As a child, how often did you play by yourself?

1. Nearly all the time.   ....

2. Very often.   ....

3. Often.   ....

4. Not often. ....
5. Hardly at all. ....

23. If there were a General Election tomorrow, for which of the following parties would you vote, assuming you were allowed to vote, and that a candidate for each party were standing? How strongly do you subscribe to the principles for which this party stands? In answering please use the following code: 1 = completely: 2 = very strongly; 3 = strongly; 4 = not strongly; 5 = do not really subscribe to them.

	Would vote for	Strength of subscription to principles
1. British National.	....	....
2. Communist.	....	....
3. Conservative.	....	....
4. Labour.	....	....
5. Liberal.	....	....
6. Other:	....	....

If you have time to spare, we would like you to indicate why you have decided to work in the location you have ticked in question 10. If you also have any comments on the questionnaire, these would be appreciated.

Thank you for your co-operation.

In the second wave of questionnaires, mainly to first year students, question 9, referring to the university social structure, was omitted. The following question was included; this also encompassed the questions 4, 5 and 22.

Would you please read each of the following statements carefully and then indicate how far you agree/disagree with it by ticking the appropriate space after the statement. *Please remember that there are no right or wrong answers.* 1 = complete agreement; 2 = agreement; 3 = mixed feelings; 4 = disagree; 5 = complete disagreement; 6 does not apply. (Please *only* tick 6 when the item has no application to you; if you cannot make up your mind, please tick 3.)

	Agree			Disagree		DNA
	1	2	3	4	5	6

It is no use complaining to public officials about what is happening in society; they never pay attention to you. ....

I often think about the question: 'what is the meaning of life?' ....

To succeed in this society, you have to forget about principles. ....

Today, there are very few people you can really trust. ....

I think of myself as a scientist first, and as a member of a nation second. ....

There is nothing unpatriotic about working in a foreign country, since a scientist's first loyalty is towards science. ....

The power of trade unions should be decreased. ....

Criminals should have physical punishment inflicted on them. ....

Premarital sexual relationships are quite permissible. ....

Ideas that have no useful or practical application are of little use to man. ....

Obedience and respect for authority should be the very first requirement of a good citizen. ....

The opinions of older people should be given much more attention than they are now. ....

Many of the values my parents hold are not relevant for my present life. ....

The type of people I choose as friends are similar to the type of people my parents choose to have as their friends. ....

I feel I live two kinds of lives: one at college, the other at home. ....

Most of my friends are intensely interested in science. ....

I find it very difficult to talk to people I once liked, but who are not going (or intending to go) to university. ....

I was more successful at science subjects at school than at other subjects. ....

I tended to admire the science teacher(s) at school more than other teachers. ....

205

I prefer going to the cinema than to the theatre.

I find meeting strangers at social occasions a difficult thing to carry off well.

I dislike formal occasions because of the rules of proper behaviour.

I prefer to be with a large group of people I know, than with one or two close friends.

I was probably one of the most popular pupils at school.

As a child, I liked my mother much more than my father.

I was allowed to decide to go out by myself, say to the cinema, park or bus/train journeys, before I was ten years old.

When I was a child, if I wanted to know why I should do something my parents would generally give me a reasonable explanation.

My parents seldom employed physical punishments against me if I misbehaved.

I often played by myself as a child.

As a child, I could have played with other children, but I often preferred to play by myself.

Note: Question differed for graduates and third-year undergraduates. In the above questionnaire, the version sent to the latter is given. For the former students, the options were suitably amended to eliminate the higher degree choices.

Index

For Product Safety Concerns and Information please contact our EU
representative GPSR@taylorandfrancis.com Taylor & Francis Verlag GmbH,
Kaufingerstraße 24, 80331 München, Germany

Printed and bound by CPI Group (UK) Ltd, Croydon, CR0 4YY
08/05/2025
01864462-0001